Lecture Notes in Statistics 77

Edited by J. Berger, S. Fienberg, J. Gani, K. Krickeberg, I. Olkin, and B. Singer

Enno Mammen

When Does Bootstrap Work?

Asymptotic Results and Simulations

Springer-Verlag
New York Berlin Heidelberg London Paris
Tokyo Hong Kong Barcelona Budapest

Enno Mammen
Institut für Angewandte Mathematik
Universität Heidelberg
Im Neuenheimer Feld 294
W-6900 Heidelberg
Germany

Mathematics Subject Classifications (1991): 62G09, 62G05, 62G07, 62J05

Library of Congress Cataloging-in-Publication Data
Mammen, E. (Enno)
When does bootstrap work? : asymptotic results and simulations /
E. Mammen.
p. cm. — (Lecture notes in statistics ; 77)
Includes bibliographical references.
ISBN 0-387-97867-4. — ISBN 3-540-97867-4
1. Bootstrap (Statistics) I. Title. II. Series: Lecture notes
in statistics (Springer-Verlag) ; v. 77.
QA276.8.M36 1992
519.5'44—dc20 92-16385

Printed on acid-free paper.

Camera ready copy provided by the author.
Printed and bound by Edwards Brothers, Inc., Ann Arbor, MI.
Printed in the United States of America.

9 8 7 6 5 4 3 2 1

ISBN 0-387-97867-4 Springer-Verlag New York Berlin Heidelberg
ISBN 3-540-97867-4 Springer-Verlag Berlin Heidelberg New York

Contents

Chapter 0. Introduction.

In these notes some results are presented for the asymptotic behavior of the bootstrap procedure. Bootstrap is a procedure for estimating (approximating) the distribution of a statistic. It is based on resampling and simulations. It was been introduced in Efron (1979) and in the last decade it has been discussed for a wide variety of statistical problems. Introductory are the articles Efron and Gong (1983) and Efron and Tibshirani (1986) and the book Helmers (1991b). Many applications of bootstrap are discussed in Efron (1982). Survey articles are Beran (1984b), Hinkley (1988), and Diciccio and Romano (1988a). For many classical decision problems (testing and estimation problems, prediction, construction of confidence regions) bootstrap has been compared with classical approximations based on mathematical limit theorems and expansions (for instance normal approximations, empirical Edgeworth expansions) (see for instance Bretagnolle (1983) and Beran (1982, 1984a, 1987, 1988), Abramovitch and Singh (1985), and Hall (1986a, 1988)). An asymptotic treatment of bootstrap is contained in the book Beran and Ducharme (1991). A detailed analysis of bootstrap based on higher - order Edgeworth expansions has been carried out in the book Hall (1992). Recent publications on bootstrap can also be found in the conference volumes LePage and Billard (1992) and Joeckel, Rothe, and Sendler (1992).

We will consider the application of bootstrap in three contexts : estimation of smooth functionals, nonparametric curve estimation, and linear models. We do not attempt a complete description of bootstrap in these areas. Rather we want to study and to clarify in some examples the conditions under which one can expect that bootstrap works satisfactorily.

The setting in which bootstrap can be applied can be described abstractly as follows. Given are n i. i. d. observations $X_1, \dots, X_n$ with common distribution P and with values in some measurable space $\mathcal{X}$. Furthermore, for a set $\mathcal{P}$ of probability measures containing P, a functional V_n is given which maps $\mathcal{X}^n \times \mathcal{P}$ into some measurable space $\mathcal{Y}$. For Q in $\mathcal{P}$ define the distribution

$$L_n(Q) = \mathcal{L}(V_n(Z_1, \dots, Z_n ; Q) : Z_1, \dots, Z_n \text{ are independent random variables each distributed according to } Q) .$$

The bootstrap is a procedure for estimating the distribution of $V_n(X_1, \dots, X_n ; P)$, i.e.

the distribution $L_n(P)$. Namely, let $\widetilde{P}_n$ be an estimate of P, then the bootstrap estimate of $L_n(P)$ is simply $L_n(\widetilde{P}_n)$.

Let us give two examples: a bootstrap confidence interval for a statistical functional and parametric bootstrap. Consider first the problem of constructing a confidence interval for a real - valued statistical functional T(P). The confidence interval can be based on the estimate $T(\widehat{P}_n)$, where $\widehat{P}_n$ denotes the empirical distribution $\widehat{P}_n(A) = n^{-1} \Sigma_{i=1}^n 1(X_i \in A)$. Put $V_n(X_1 , \dots , X_n ; P) = T(\widehat{P}_n) - T(P)$. Then an approximate $(1-\alpha)$ confidence interval can be based on the $\alpha/2$ and $1-\alpha/2$ quantile of $L_n(\widehat{P}_n)$.

For the parametric bootstrap consider the case that $\mathcal{P}$ is a parametric family $\{P_\theta : \theta \in \Theta\}$ and that operational characteristics of an estimator $\widehat{\theta}_n = \widehat{\theta}_n(X_1 , \dots , X_n)$ are of interest. Then one may put $V_n(X_1, \dots , X_n ; P_\theta) = \widehat{\theta}_n - \theta$ and one may estimate $L_n(P_\theta)$ by $L_n(P_{\widetilde{\theta}_n})$ where $\widetilde{\theta}_n$ is a consistent estimate of the underlying θ (often $\widetilde{\theta}_n = \widehat{\theta}_n$) .

It is important to mention that bootstrap can also be used in other settings than i. i. d. models. In Chapter 1 we consider bootstrap in models with independent observations which are not identically distributed. Furthermore, some proposals have been made in the literature for dependent observations. For autoregressive processes bootstrap procedures which rely on the i. i. d. structure of the error variables have been studied in Bose (1988) and Basawa, Mallik, McCormick, and Taylor (1989). For stationary sequences Kreiss (1988) considers bootstrap in a approximating model of higher - order autoregressive processes. Resampling of blocks of consecutive observations has been proposed in Künsch (1989). A bootstrap procedure which mimics structures in the frequency domain can be found in Franke and Härdle(1990). For bootstrap of nonparametric nonlinear time series see Franke (1990).

In Chapters 1, …, 4 of these notes we will study the range of validity of bootstrap. This will be done in Chapter 1 for a linear statistical functional $T_n(Q) = \int g_n \, dQ$ and in Chapters 2, …, 4 in the context of nonparametric curve estimation. Note that bootstrap needs two conditions to work:

(i) *"Continuity" of the functional* $L_n : Q \to L_n(Q)$ *in a neighborhood of* P *in* $\mathcal{P}$.

(ii) *Consistency of the estimate* $\widetilde{P}_n$.

In the first chapter we will show that in the case of a linear statistical functional $T_n(Q) = \int g_n \, d\, Q$ these conditions are equivalent to a Lindeberg condition of $g_n(X_1)$ and therefore also to the asymptotic normality of $T_n(\widehat{P}_n) - T_n(P)$. This equivalence of asymptotic normality and of consistency of bootstrap does not hold in general for arbitrary functionals T. This has been pointed out by Beran (1982) using the Hodges estimate, which is defined for $b \neq 1$ by:

$$V_n(X_1, \dots , X_n ; P) = \begin{cases} b\,\overline{X}_n - E_P(X) & \text{if } |\overline{X}_n| < n^{-1/4} \\ \overline{X}_n - E_P(X) & \text{else,} \end{cases}$$

where $\overline{X}_n = n^{-1} \Sigma_{i=1}^{n} X_i$ is the arithmetic mean. Here it can easily be shown that the bootstrap estimate of $L_n(P)$ (the distribution of $V_n(X_1 , \dots , X_n ; P)$) is not consistent for measures P with mean $E_P(X) = 0$. Note that if the variance of X_i is finite one gets for large n

$$L_n(P) \approx N(0 , n^{-1} \; var_P(X)) \;\; \text{if } E_P(X) \neq 0$$

and

$$L_n(P) \approx N(0 , n^{-1} \; b^2 \, var_P(X)) \;\; \text{if } E_P(X) = 0.$$

This implies that L_n is not "asymptotically continuous" at measures P with mean 0.

Another example will be given in Chapter 2, where the L^2-distance between a nonparametric kernel regression estimator and a parametric regression estimator is proposed as goodness - of - fit test statistic of the parametric regression model. It is shown that bootstrap does not estimate the distribution of this test statistic consistently on the parametric null hypothesis. Thus bootstrap cannot be used here for an approximate determination of critical values. Here too the test statistic is asymptotically normal. We present an alternative resampling procedure ("wild bootstrap"), which does work. In Chapter 3 a test statistic is considered which is statistically and mathematically related to the test statistic discussed in Chapter 2. It will be shown that bootstrap works here for a choice of V_n which is *not* of the form $T_n(\widetilde{P}_n) - T_n(P)$. This highlights the point that the appropriate choice of V_n (and of $\widetilde{P}_n$) is crucial for consistency of bootstrap. Other examples where bootstrap does not work can be found in Beran and Srivastava (1985), Basawa, Mallik, McCormick, Reeves, and Taylor (1991), and Dümbgen (1991b).

Explicit calculation of the bootstrap estimate $L_n(\widetilde{P}_n)$ is almost always prohibitively complicated. It is therefore usual to approximate the estimate numerically using resampling and simulations. The procedure is as follows.

BOOTSTRAP RESAMPLING:

STEP 1. *Generate a sample* $X_1^*, \ldots, X_n^*$ *of i. i. d. (pseudo) random variables with distribution* $\widetilde{P}_n$.

STEP 2. *Calculate* $V_n^* = V_n(X_1^*, \ldots, X_n^*; \widetilde{P}_n)$.

STEP 3. *Repeat the first two steps, say* M *times, obtaining the values* $V_{n,1}^*, \ldots, V_{n,M}^*$.

STEP 4. *Approximate* $L_n(\widetilde{P}_n)$ *by the empirical distribution of the (pseudo) sample* $V_{n,1}^*, \ldots, V_{n,M}^*$.

In these notes all theoretical results are stated for the bootstrap estimate $L_n(\widetilde{P}_n)$ (i.e. $M = \infty$). We will not address algorithmic problems. An appropriate choice of M has been discussed for instance in Hall (1986b) and Hall and Titterington (1989). Approximations for the bootstrap estimate which avoid Monte Carlo calculations have been proposed in Davison and Hinkley (1988) and DiCiccio and Efron(1990) (see also Efron, 1990). For reducing the number of bootstrap replications importance sampling has been proposed in Johns (1988), Hall (1991), and Do and Hall (1991).

For complicated functionals V_n and for large data sets the bootstrap algorithm may be very computer intensive. This may be considered as not very disadvantageous when enough computer power is available or when no alternative procedure is known for which faster algorithms exist. Examples for the second case are some problems in nonparametric curve estimation, where asymptotics leads to approximations based on functionals of Gaussian processes. Often the asymptotic distributions of these functionals cannot be calculated explicitly or explicit approximations are so poor that, typically, they cannot be used in practice for moderate sample sizes. For instance, we think here of choosing L_p - distances between a curve estimate and the underlying curve ($p = 1, 2$, or ∞) as pivots V_n as is be done in Chapters 2 and 3. Other examples are shape characteristics of curve estimates. In particular, in Chapter 4 we consider the number of modes of kernel regression estimates. There we discuss a test procedure proposed by Silverman (1981) for the determination of the number of modes of a density. This is an example where bootstrap works quite well and is at present the only satisfactory method for determing critical values.

Bootstrap has also been applied in models where the classical asymptotic distribution approximations are very easily to compute. Then one may ask why one should apply bootstrap in these models and what are here the advantages of bootstrap. We consider two answers to this question.

A first answer can be given using higher - order (Edgeworth) expansions. For linear statistics this has been done by Singh (1981) and Bickel and Freedman (1980). For the more general class of statistical functionals which are functions of vector means Hall (1986a, 1988, 1992) has shown that bootstrap of the studentized functional has a better rate of convergence than the normal approximation. For a simple explanation of this result see Hartigan (1986). Extensive discussion of bootstrap using higher - order expansions can be also found in Beran (1982, 1984a, 1987, 1988). In Chapter 5 we give a simple proof for the higher - order accuracy of the bootstrap estimate. This will be done for arbitrary smooth statistical functionals. The idea of the proof is to make a direct comparison of expectations of smooth functions in the "bootstrap world" and in the "real world". The proof does not use Edgeworth expansions.

A second answer to our question has been given by Bickel and Freedman (1983) in the context of a linear model. They show that bootstrapping the least squares estimator works under essentially weaker conditions than those which are necessary for the classical normal approximation. This type of result is of much greater significance for the applied statistician than the first answer given above. The first answer merely says that if the classical normal approximation is good, then the bootstrap is even better. But this additional accuracy is not needed in many applications. The second answer, by contrast, says that we can apply the bootstrap in cases in which the normal approximation breaks down. This is a real gain. Bickel and Freedman's approach allows for the presence of individual influential data points (leverage points) and it allows the dimension of the model to depend on n. (This is appropriate for many applications in which the dimension is not small compared with the sample size.) A short description of their results and an asymptotic treatment of F-tests is given in Chapter 6. In Chapter 7 the results of Bickel and Freedman are generalized to M - estimates. This shows that bootstrap works also for more complicated estimators, even if the dimension is large and the data contain leverage points. An asymptotic approach, where the dimension and other model characteristics may depend on the sample size n, is very natural because it clarifies what is needed for a good approximation for finite sample size n. Nevertheless, it may be very difficult to use this approach for other models. First, this concerns the technical methods which are needed for such an asymptotic treatment. Furthermore, typically it is not a priori clear in which sequence of model classes a model should be imbeddded which is given for *one* fixed sample size n (Consider for instance nonlinear models with increasing dimension). Therefore we have used this approach in these lecture notes only for the relatively simple case of linear models. For loglinear models asymptotics of this type have been carried out in Ehm (1987, 1991).

Bootstrap behaves quite differently in high dimensional linear random design models as in models with fixed design. This will be explained in Chapter 8. Here bootstrap is based on resampling from the tuples consisting of the design vector and the observation. The bootstrap has to mimic here a more complex stochastic structure than in the fixed design model with one dimensional i. i. d. errors. In Chapter 8 we compare the range of validity of bootstrap, wild bootstrap (see also Chapters 1 and 2), and normal approximations for least squares estimators. Furthermore we find that for models with increasing dimension bootstrap and wild bootstrap have differing rates of convergence. This explains the different performance of these procedures in a simulation study presented in Chapter 8. Note that the usual asymptotic approach, in which the dimension p is fixed, does not give any explanation of these differences, because then all procedures have the same rate of convergence.

Acknowledgments. I would like to thank W. Ehm and D. W. Müller for valuable comments and discussions. In particular, they proposed to use an asymptotic approach where the model may depend on the number of observations (see Chapters 6 and 7). I wish also to thank H. Dinges, B. Ferebee, and G. Kersting for the opportunity to give a course on bootstrap at the university of Frankfurt and for interrupting this course by many interesting remarks and discussions. In particular, Chapters 1 and 5 are motivated by discussions with them. B. Ferebee and G. Sawitzki have helped me with the english grammar. They have made a number of suggestions concerning the presentation of these notes. My interest in bootstrap methods has been stimulated by talks of R. Beran and conversations with him during his several stays at the university of Heidelberg. I would also like to thank P. Hall for kindly sending me a copy of his book manuscript on bootstrap and Edgeworth expansions and for providing help for the improving of the quality of some figures in these notes. Chapters 2 and 4 are based on joint works with W. Härdle and with N. Fisher and S. Marron. The simulations of Chapter 8 have been programmed by M. Suckut. These lecture notes are a revised version of a Habilitationsschrift written at the Universität Heidelberg. This work has been supported by the Sonderforschungsbereich 123 "Stochastische Mathematische Modelle", Universität Heidelberg.

Chapter 1. Bootstrap and asymptotic normality.

1. Introduction. In this chapter consistency of bootstrap is compared with asymptotic normality. This is done for linear statistics of n i. i. d. observations. It is shown that bootstrap works asymptotically under the same assumptions as a normal approximation with estimated variance (Theorem 1). This result is extended to the case of independent but not necessarily identically distributed observations (Theorem 2). Furthermore, bootstrap with a Poisson random sample size is considered. This bootstrap procedure is a special case of a class of resampling plans called wild bootstrap which have been proposed for the non i. i. d. case. We show that bootstrap works as long as the same holds for wild bootstrap (Theorem 3). The poissonisation is also a central tool for proving that consistency of bootstrap implies asymptotic normality.

2. Bootstrapping linear functionals. The i. i. d. case. Let us first consider samples $X_n = (X_{n,1}, \dots, X_{n,n})$ of n i. i. d. variables with unknown distribution P_n. We study the following bootstrap procedure for estimating $P(T_n(\widehat{P}_n) - T_n(P_n) \le t)$, where $\widehat{P}_n$ is the empirical distribution based on the sample. The bootstrap estimate is $P^*(T_n(\widehat{P}_n^*) - T_n(\widehat{P}_n) \le t)$. Here P^* denotes the conditional law $P(\dots|X_n)$. $\widehat{P}_n^*$ is the empirical distribution of a bootstrap sample $X_n^* = (X_{n,1}^*, \dots, X_{n,n}^*)$; that is, given X_n, the bootstrap sample X_n^* is an i. i. d. sample with distribution $\widehat{P}_n$. For fixed $P_n = P$ and "smooth" $T_n = T$ an elegant proof of the consistency of bootstrap can be found in Gill (1989) for asymptotically linear functionals (see also Arcones and Giné, 1990). In Chapter 5 a simple treatment of higher - order properties is given for smooth functionals.

Here we consider the case of a linear functional $T_n(Q) = \int g_n(x)\ Q(dx)$. We write $\widehat{T}_n = T_n(\widehat{P}_n)$. For fixed $T_n = T$ and $P_n = P$ it is known that bootstrap is (weakly) consistent if and only if $g(X_{n,1})$ belongs to the domain of attraction of the normal law (see Hall (1990d), Giné and Zinn (1989), Csörgö and Mason (1989), Athreya (1987), and Knight (1989)). Necessary and sufficient conditions for the (infinite - dimensional) case of bootstrapping empirical measures can be found in Giné and Zinn (1990). In our approach we allow everything to depend on n. There are two reasons for doing this: first, as a first step into the direction of studying an arbitrary sequence of statistics $\widehat{T}_n$; second, because the nature of needed asssumptions can be better understood. In particular, this makes it easier to judge if the asymptotic results apply for a finite sample size n. We will show in the following theorem that asymptotic normality is necessary and sufficient for consistency of bootstrap. The proof will be based on the following simple argument. If bootstrap works, the bootstrap estimate of the distribution of the standardized functional must be asymptotically equivalent to a sequence of *non - random* distributions. We will show that for this asymptotic equivalence it is necessary and sufficient that the absolute maximal summand $\max_{1 \le i \le n} | g_n(X_{n,i})|$ is of smaller order than the sum $\hat{T}_n$. This Lindeberg - type condition implies the asymptotic normality of $\hat{T}_n$. Note that we do not have to treat the two cases $E\, g_n^2(X_{n,i}) < \infty$ and $E\, g_n^2(X_{n,i}) = \infty$ separately.

THEOREM 1. *Consider a sequence* $X_{n,1}, \ldots, X_{n,n}$ *of i. i. d. variables with distribution* P_n. *For a function* g_n *consider* $\hat{T}_n = T_n(\hat{P}_n) = \frac{1}{n}\sum_{i=1}^{n} g_n(X_{n,i})$. *Consider a bootstrap sample* $X^*_{n,1}, \ldots, X^*_{n,n}$ *with empirical distribution* $\hat{P}^*_n$. *Denote* $\hat{T}^*_n = T_n(\hat{P}^*_n)$. *Then for every sequence* t_n *the following assertions are equivalent:*

(i) $\hat{T}_n$ *is asymptotically normal: There exist* σ_n *with*

$$(2.1) \qquad d_\infty (\mathcal{L}(\hat{T}_n - t_n) , N(0,\sigma_n^2)) \to 0.$$

(ii) *The normal approximation with estimated variance works:*

$$(2.2) \qquad d_\infty (\mathcal{L}(\hat{T}_n - t_n) , N(0, \hat{S}_n^2)) \to 0 \quad (\textit{in probability}),$$

where $\hat{S}_n^2 = \frac{1}{n^2} \sum_{i=1}^{n} \left(g_n(X_{n,i}) - \hat{T}_n\right)^2.$

(iii) *Bootstrap works:*

$$(2.3) \qquad d_\infty (\mathcal{L}(\hat{T}_n - t_n) , \mathcal{L}^* (\hat{T}^*_n - \hat{T}_n)) \to 0 \ (\textit{in probability}).$$

Here d_∞ *denotes the Kolmogorov distance and* $\mathcal{L}^*(\ldots)$ *is the conditional law* $\mathcal{L}(\ldots | X_{n,1},\ldots, X_{n,n})$. *If* (i), (ii) *or* (iii) *hold then* t_n *can be chosen as mean of truncated variables* $\mu + E\,(g_n(X_{n,i}) - \mu)\ 1(|g_n(X_{n,i}) - \mu| \le n\,\sigma_n)$, *where* μ *is a median of the distribution of* $g_n(X_{n,1})$.

Under the additional assumption of

$$\sigma_n^{-1}\, E\, |g_n(X_{n,i}) - t_n|\, 1(|g_n(X_{n,i}) - t_n| > n\, \sigma_n) \to 0,$$

the sequence t_n can be replaced by $E(\widehat{T}_n) = T_n(P_n)$. Then according to the theorem bootstrap works as an estimate of $P(T_n(\widehat{P}_n) - T_n(P_n) \leq t)$ if and only if the same holds for the classical approximation $N(0,\widehat{S}_n^2)$. It can also be shown that, under the conditions of the theorem, bootstrap of the studentized functional works.

For $g_n(x) = x$ the sequence t_n can be chosen as $T(P_n^{'})$ where $P_n^{'}$ is the distribution of $X_{n,i}^{'} = \mu + (X_{n,i} - \mu)\, 1(|X_{n,i} - \mu| \leq n\, \sigma_n)$ where now μ is a median of the distribution of $X_{n,1}$. If (i), (ii) or (iii) hold then one can show that $P(X_{n,i}^{'} = X_{n,i}$ for $1 \leq i \leq n) \to 1$, i. e. asymptotically one cannot distinguish if a sample comes from P_n or $P_n^{'}$.

A simple example where the conditions of the theorem are not fulfilled is given by the following triangular array:

$$g_n(X_{n,i}) = \begin{cases} 0 & \text{with probability } 1 - \frac{1}{n} \\ 1 & \text{with probability } \frac{1}{n}. \end{cases}$$

Then $\widehat{T}_n$ converges weakly to a Poisson distribution with mean 1 . This does not hold in the bootstrap world, because for instance

$$P(\mathcal{L}^*(\widehat{T}_n^* - \widehat{T}_n) = \delta_0) \geq P(g_n(X_{n,1}) = \ldots = g_n(X_{n,n}) = 0)$$

$$= (1 - n^{-1})^n \to e^{-1} > 0$$

for $n \to \infty$.

We get another example where the conditions of the theorem are not fulfilled by taking fixed $g_n = g$ and P_n = the distribution of $g(X_{n,1}) = P$, where P belongs to the domain of attraction of a stable law (see Athreya (1987) , Knight (1989)).

3. Bootstrapping smooth functionals. Let us shortly also discuss here nonlinear functionals T. In case of nonlinear functionals T, *typically* local asymptotic linearity seems to be necessary for consistency of bootstrap of T if V_n (see Chapter 0.

Introduction) is chosen as $V_n = T(\hat{P}_n) - T(P_n)$. This has been pointed out by van Zwet (1989) in an analysis based on the Hoeffding decomposition of a statistic $\hat{T}_n$:

$$\hat{T}_n = \sum_{I \subset \{1, \dots, n\}} \hat{T}_{n,I},$$

where

$$\hat{T}_{n,I} = \sum_{J \subset I} (-1)^{\# I - \# J} \quad E(\hat{T}_n \mid X_{n,i} : i \in J).$$

If the linear term $\sum_{i=1}^{n} \hat{T}_{n,\{i\}}$ is of lower order than $\hat{T}_n$, bootstrap cannot work because it does not estimate consistently the variance of higher - order terms $\sum_{\# I = k} \hat{T}_{n,I}$ for $k > 1$ (see also Sen, 1988b). Consider for instance a U-statistic $\hat{T}_n = \frac{1}{n} \sum_{i \neq j} W(X_i, X_j)$ which is "pure" (i.e. $E(W(X_i, X_j) \mid X_i) = E(W(X_i, X_j) \mid X_j) = 0$ for $i \neq j$). Here we have written $X_i = X_{n,i}$ and supposed that $P = P_n$ does not depend on n. Then $E\, var^*(\hat{T}_n^*) - 3\, var\, (\hat{T}_n) \to 0$ under suitable moment conditions (see the lemma and the following discussion in Chapter 2 Section 2). The bootstrap and the jackknife estimate of variance have also been studied in an approach based on the Hoeffding decomposition in Efron and Stein (1981), Efron (1982), and Karlin and Rinott (1982). For a discussion of the Hoeffding decomposition see also van Zwet (1984).

In case of nonlinear functionals T, asymptotic normality does not imply consistency of bootstrap. A simple example is the Hodges estimate (Beran (1982), see the introduction). Other examples are pure U - statistics $\hat{T}_n = \sum_{i \neq j} W_n(X_i, X_j)$ where the kernel W_n may depend on n. For simple conditions under which U-statistics are asymptotically normal see de Jong (1987). An example is $\hat{T}_n = n\sqrt{h_n} \int (\hat{f}_{h_n} - K_{h_n} * f)^2\, dx$ where $f = F'$ is the underlying density and where $\hat{f}_h$ is a kernel estimate with kernel $K_h(u) = 1/h \ K(u/h)$. Note that $\hat{T}_n$ is asymptotically a pure U-statistic:

$$\hat{T}_n = h_n^{-1/2} (K_{h_n} * K_{h_n})(0)$$

$$+ \sum_{i \neq j} U_n(X_i, X_j) - E\ (U_n(X_i, X_j) \mid X_i)$$

$$- E(U_n(X_i, X_j) \mid X_j) + E\, U_n(X_i, X_j) + o_P(1),$$

where $U_n(X_i, X_j) = h_n^{1/2}\, n^{-1}\ (K_{h_n} * K_{h_n})(X_i - X_j)$.

The asymptotic normality of $\widehat{T}_n$ for $h_n \to 0$ has been proved by Hall (1984). Bootstrap does not work here. Indeed, one can show that : $\mathrm{var}^*\left(\widehat{T}_n^*\right) - 3\ \mathrm{var}\ (\widehat{T}_n) \to 0$ (in probability), i.e. the bootstrap variance estimate is not consistent. This example for inconsistency of bootstrap may seem artificial, because the knowledge of the underlying density is used in the construction of the statistic, but not in the resampling step of the bootstrap procedure. A mathematically related example which makes more sense statistically can be found in Chapter 2. There the L^2-distance between a nonparametric kernel regression estimator and a parametric regression estimator is proposed as a goodness - of - fit test statistic of a parametric regression model. It is shown that bootstrap does not estimate the distribution of the test statistic consistently on the hypothesis. The test statistic is asymptotically equivalent to a pure U-statistic and asymptotically normal.

4. Bootstrap and wild bootstrap in non i. i. d. models. Although in the bootstrap procedure an i. i. d. model is mimicked, bootstrap works also for models with independent but not necessarily identically distributed observations. This has been observed first in Liu (1988) and Liu and Singh (1991). As in Section 2 we consider now linear functionals T_n. In the next theorem we show that again bootstrap works as long as the normal approximation with estimated variance is asymptotically consistent. However, now this is not equivalent to asymptotic normality.

THEOREM 2. *Consider a sequence* $X_{n,1}, \ldots, X_{n,n}$ *of independent random variables with distribution* $P_{n,i}$. *For a function* g_n *we define* $\widehat{T}_n$ and $\widehat{T}_n^*$ *as in Theorem 1. Then for every sequence* t_n *the following assertions are equivalent:*

(i) *There exist* σ_n *such that for every* $\varepsilon > 0$

$$\text{(4.1)} \qquad \sup_{1\le i\le n} P\left(\left|\frac{g_n(X_{n,i}) - t_n}{n\,\sigma_n}\right| \ge \varepsilon\right) \to 0$$

(asymptotic negligibility),

(4.2)
$$\sum_{i=1}^{n}\left(E\,\frac{g_n(X_{n,i}) - t_n}{n\,\sigma_n}\,1\left(\left|\frac{g_n(X_{n,i}) - t_n}{n\,\sigma_n}\right| \le \varepsilon\right)\right)^2 \to 0,$$

and such that

$$d_\infty\left(\mathcal{L}\left(\hat{T}_n - t_n\right), N(0,\sigma_n^2)\right) \to 0.$$

(ii) *The normal approximation with estimated variance works:*

$$d_\infty\left(\mathcal{L}(\hat{T}_n - t_n), N(0, \hat{S}_n^2)\right) \to 0 \quad \textit{(in probability)},$$

where again $\hat{S}_n^2 = \frac{1}{n^2}\sum_{i=1}^{n}\left(g_n(X_{n,i}) - T_n(\hat{P}_n)\right)^2.$

(iii) *Bootstrap works:*

$$d_\infty\left(\mathcal{L}(\hat{T}_n - t_n), \mathcal{L}^*(\hat{T}_n^* - \hat{T}_n)\right) \to 0 \quad \textit{(in probability)}.$$

It is clear that asymptotic normality, i. e. $d_\infty(\mathcal{L}(\hat{T}_n - t_n), N(0,\sigma_n^2)) \to 0$, is now not sufficient for consistency of bootstrap. Consider for instance the case that $X_{n,1} = \ldots = X_{n,n-1} = 0$ (a. s.) and that $\mathcal{L}(X_{n,n}) = N(0, 1)$. Such cases are excluded by (4.1). The role of (4.2) can also be explained easily. As in Theorem 1 we have not assumed that $T_n(P_{n,i})$ is finite. Without loss of generality we consider again the case $g_n(x) = x$ and we define $P'_{n,i}$ as P'_n in the discusssion after Theorem 1. Then (4.2) says that the differences between $T_n(P'_{n,i})$ for different i are not too large. This is necessary for consistency of the normal approximation and of the bootstrap procedure Otherwise the differences between $T_n(P'_{n,i})$ would be interpreted as stochastic fluctuations by the variance estimate $\hat{S}_n^2$ and by bootstrap. This would lead to overestimation of variance.

For understanding why bootstrap works also for non i. i. d. observations let us introduce another resampling plan. Consider a Poisson variable N with EN = n which is independent of $(X_{n,1}, \ldots, X_{n,n})$. Given $X_{n,1}, \ldots, X_{n,n}$ and N = k we generate (conditionally) independent variables $X^*_{n,1}, \ldots, X^*_{n,k}$ with conditional distribution $\hat{P}_n$. We write $Y_{n,i} = n^{-1} g_n(X_{n,i})$ and $Y^*_{n,i} = n^{-1} g_n(X^*_{n,i})$. We consider

$$\widehat{T}_n^{*,P} = Y_{n,1}^* + \ldots + Y_{n,k}^*.$$

Then $\mathcal{L}^*(\widehat{T}_n^{*,P} - \frac{N}{n}\widehat{T}_n)$ can be used as an estimate of $\mathcal{L}(\widehat{T}_n - t_n)$. We show that this estimate works under the same assumptions as bootstrap.

THEOREM 3. *Under the assumptions of Theorem 2 the following assertions are equivalent for every sequence* t_n:

(i) *Bootstrap works:*

$$d_\infty(\mathcal{L}(\widehat{T}_n - t_n), \ \mathcal{L}^*(\widehat{T}_n^* - \widehat{T}_n)) \to 0 \qquad \textit{(in probability)},$$

(ii) *Bootstrap with Poisson sample size works:*

$$d_\infty(\mathcal{L}(\widehat{T}_n - t_n), \ \mathcal{L}^*(\widehat{T}_n^{*,P} - \frac{N}{n}\widehat{T}_n)) \to 0 \qquad \textit{(in probability)}.$$

We have introduced bootstrap resampling with Poisson random sample size, because this resampling plan has the following nice interpretation. Denote

$$N_j = \# \{ 1 \le i \le N: \ X_{n,i}^* = X_{n,j} \}.$$

Suppose that the $X_{n,1}, \ldots, X_{n,n}$ are pairwise different (a.s.). Then the N_j's are independent Poisson variables with $E\,N_j = 1$ (a.s.). Furthermore

$$\begin{aligned} \widehat{T}_n^{*,P} - \frac{N}{n}\widehat{T}_n &= \sum_{j=1}^{n} N_j\,(Y_{n,j} - \widehat{T}_n / n) \\ &= \textstyle\sum_{j=1}^{n} (N_j - 1)\,(Y_{n,j} - \widehat{T}_n / n) \qquad \text{(a.s.).} \end{aligned}$$

We write this as

$$\widehat{T}_n^W = \textstyle\sum_{j=1}^{n} Y_{n,j}^W, \tag{4.3}$$

where $Y_{n,j}^{W} = (N_j - 1)\,(Y_{n,j} - \hat{T}_n / n)$.

Because of $E(N_j - 1) = 0$ and $E(N_j - 1)^2 = E(N_j - 1)^3 = 1$ one gets for the conditional expectations of $Y_{n,j}^{W}$

$$(4.4) \qquad E^* \, Y_{n,j}^{W} = 0,$$

$$(4.5) \qquad E^*(Y_{n,j}^{W})^2 = (Y_{n,j} - \hat{T}_n / n)^2,$$

and

$$(4.6) \qquad E^*(Y_{n,j}^{W})^3 = (Y_{n,j} - \hat{T}_n / n)^3.$$

Resampling plans with (4.3), ... , (4.6) have been introduced in Wu (1986) (see also Beran, 1986). They have been called in Härdle and Mammen (1990) *wild bootstrap* for the following reason. $\mathcal{L}^*(Y_{n,j}^{W})$ could be interpreted as an estimate of $\mathcal{L}(Y_{n,j} - E\,Y_{n,j})$. This estimate is based on only *one* residual $Y_{n,j} - \hat{T}_n / n$. The seemingly daring, even reckless, nature of this enterprise is the reason for the name *wild bootstrap*. Condition (4.6) has been introduced by Liu (1988) and Härdle and Mammen (1990) to improve the rate of convergence of the wild bootstrap estimate. Consistency of wild bootstrap has been proved by Liu (1988) for a special model, who also gives some heuristic arguments that the wild bootstrap estimate of the distribution of studentized estimators converges with rate 1/n. Wild bootstrap in non i. i. d. models has also been studied in Liu and Singh (1989, 1991). Application of wild bootstrap to nonparametric curve estimation can be found in Chapter 2. Wild bootstrap in linear models is discussed in Chapter 8. Here we consider the following class of wild bootstrap procedures:

WILD BOOTSTRAP:

(4.7) *Choose a distribution* Q *with* $E(Z \mid Q) = 0$ *and* $E(Z^2 \mid Q) = 1$.

(4.8) *Generate i. i. d. variables* $Z_1^W, \ldots, Z_n^W$ *with distribution* Q *and put*

$$\hat{T}_n^W = \sum_{j=1}^{n} Y_{n,j}^{W},$$

where $Y_{n,j}^{W} = Z_j^W\,(Y_{n,j} - \hat{T}_n / n)$.

(4.9) *Estimate* $\mathcal{L}(\hat{T}_n - t_n)$ *by* $\mathcal{L}^*(\hat{T}_n^W)$.

The next theorem shows that this class of wild bootstrap estimates works under the same conditions as bootstrap.

THEOREM 4. *Consider a wild bootstrap procedure of the form* (4.7), ..., (4.9) *with fixed* Q. *Then under the assumptions of Theorem 2 the following assertions are equivalent for every sequence* t_n:

(i) *Bootstrap works:*

$$d_\infty(\mathcal{L}(\hat{T}_n - t_n),\ \mathcal{L}^*(\hat{T}_n^* - \hat{T}_n)) \to 0 \qquad \textit{(in probability)},$$

(ii) *Wild bootstrap works:*

$$d_\infty(\mathcal{L}(\hat{T}_n - t_n),\ \mathcal{L}^*(\hat{T}_n^W)) \to 0 \qquad \textit{(in probability)}.$$

Note that this class of resampling procedures contains as special cases bootstrap with Poisson sample size ($Q = \mathcal{L}(N_j - 1)$) and the normal approximation $N(0, S_n^2)$ ($Q = N(0,1)$). Let us mention also three other choices of Q with $E(Z^3 \mid Q) = 1$. If one assumes that Q puts mass only on two points then $Z_i^W = -(\sqrt{5} - 1)/2$ with probability $(\sqrt{5}+1)/(2\sqrt{5})$ and $Z_i = (\sqrt{5} + 1)/2$ with probability $1 - (\sqrt{5}+1)/(2\sqrt{5})$. In another construction one may put $Z_i^W = V_i/\sqrt{2} + (V_i^2 - 1)/2$ where the V_i's are independent N(0,1) - distributed variables. In all simulations in this lecture notes Z_i^W has be chosen as $Z_i^W = (\delta_1 + V_{i,1}/\sqrt{2})(\delta_2 + V_{i,2}/\sqrt{2}) - \delta_1 \delta_2$ where the $V_{i,j}$'s are independent N(0,1) - distributed variables and where $\delta_1 = (3/4 + \sqrt{17}/12)^{1/2}$ and $\delta_2 = (3/4 - \sqrt{17}/12)^{1/2}$. For other constructions of $\hat{F}_i$ see also Liu (1988).

5. Some simulations. We use now simulations to compare bootstrap, wild bootstrap and normal approximation (with estimated variance). For samples of 20 independent observations $X_1, \ldots, X_{20}$ we consider the following models:

MODEL A. $X_1, \ldots, X_{20}$ *are i. i. d. and have a* χ^2 *distribution with 8 degrees of freedom.*

MODEL B. $X_1, \ldots, X_{20}$ *are i. i. d. and have a* χ^2 *distribution with 4 degrees of freedom.*

MODEL C. $X_1, \ldots, X_{20}$ *are i. i. d. and have a normal mixture distribution* $.5\ N(.5, (.5)^2) + .5\ N(-.5, (.7)^2)$.

MODEL D. $X_1, \ldots, X_{20}$ *are independent and* X_i *has a normal mixture distribution* $\frac{i-1}{n-1} N(\frac{n-i}{n-1}, (.5)^2) + \frac{n-i}{n-1} N(-\frac{i-1}{n-1}, (.7)^2)$.

MODEL E. $X_1, \ldots, X_{20}$ *are independent and* $X_i / [1 + 2\frac{i-1}{n-1}]$ *has a normal mixture distribution* $.5\ N(.5, (.5)^2) + .5\ N(-.5, (.7)^2)$.

MODEL F. $X_1, \ldots, X_{20}$ *are independent and* $X_i / [1 + 2\frac{i-1}{n-1}]$ *has a normal mixture distribution* $\frac{i-1}{n-1} N(\frac{n-i}{n-1}, (.5)^2) + \frac{n-i}{n-1} N(-\frac{i-1}{n-1}, (.7)^2)$.

We have run 16 000 simulations to calculate the values of u with $P(\hat{T}_n - T(P) \le u) = .01, j/20$ (for $j = 1, \ldots, 19$), and .99. T is the mean functional $T(Q) = \int x\, dQ$. Then, 10000 times, we have generated samples of 20 observations according to the Models A, ..., F. For every value of u and every sample we have calculated the bootstrap estimate, the wild bootstrap estimate, and the normal approximation. For the normal approximation the empirical variance (and not the true variance) has been used. In the bootstrap and wild bootstrap resampling 1000 replications have been used. The expected error, bias, and standard deviation of these three procedures are plotted in Figures 1, ..., 6. Every marked point in the plots corresponds to one Monte Carlo result. Because there are no large differences in the plots we present the bias and standard deviation plots only for Models A and F. Furthermore for Models A and F the standard deviations (neglecting the error in the determinbation of the u's) of the Monte Carlo calculations of the expected errors have been plotted. One sees that the standard deviations of the Monte Carlo calculations are approximately 1.5 % of the expected errors, so that if one takes into account an error of slightly smaller order in the determination of the u's one ends up with a 'standard deviation' of approximately 2%.

The simulation results look quite similarly for bootstrap, wild bootstrap, and normal approximations. The standard deviations are larger for wild bootstrap and slightly larger for bootstrap than for the normal approximation. This may be explained by the discrete nature of the bootstrap procedures. For all procedures the bias is slightly smaller than the standard deviation. Moreover, although the bias is of the same order for all procedures, the bias curves look quite differently for the normal approximation than for the bootstrap procedures: whereas for the bootstrap procedures the bias curves are nearly antisymmetric around .5, the curves are skew for the normal approximation. This may reflect the point, that the bootstrap procedures correct for skewness of $\hat{T}_n$ - T(P), however, in this example one pays for the skewness correction by getting other contributions to the bias.

These features cannot be explained by second order Edgeworth expansions and may be therefore not apparent for larger sample sizes. Indeed, suppose that for $\mathcal{L}(Z) = \mathcal{L}(\hat{T}_n - T(P))$, $\mathcal{L}^* (\hat{T}_n^* - \hat{T}_n)$, and $\mathcal{L}^*(\hat{T}_n^W)$ the formal Edgeworth expansion holds:

$$P(Z \leq x) = \Phi(\tilde{x}) + \gamma(Z)\,(1-\tilde{x}^2)\,\varphi(\tilde{x}) + O(\delta(Z)),$$

where $\tilde{x} = (x - EZ)\,var(Z)^{-1/2}$, $\gamma(Z) = (1/6)\,E\,\tilde{Z}^3$, and $\delta(Z) = E\,\tilde{Z}^4 - 3 + \gamma(Z)^2$ with $\tilde{Z} = (Z - EZ)\,var(Z)^{-1/2}$. A second order analysis based on this expansion and neglecting terms of order $o(n^{-1/2})$ suggests that the bootstrap and wild bootstrap estimates of $P(\hat{T}_n - T(P) \leq u)$ have the same asymptotic variance as the normal approximation and that they are asymptotically unbiased. However, the normal approximation has a bias (coming from the skewness term $\gamma(Z)\,(1-\tilde{x}^2)\,\varphi(\tilde{x})$) which is of the same order as the standard deviation (for more details on this arguments see also Liu and Singh, 1987). In contrast to our simulations, this would imply that the normal approximation would always have larger expected squared error than bootstrap and wild bootstrap.

The discussion based on Edgeworth expansions suggests that one can improve the accuracy of bootstrap and wild bootstrap to $o(n^{-1/2})$ if one estimates the distribution function of the *studentized* estimate $P(\hat{S}_n^{-1}(\hat{T}_n - T(P)) \leq v)$. This has been made precise in this and other contexts and in particular, in Chapter 5 it will be shown for smooth functionals. There also references to other literature can be found. In Figures 1, ...,6 the bootstrap estimate $P^*((\hat{S}_n^*)^{-1}(\hat{T}_n^* - \hat{T}_n) \leq v)$ and the wild bootstrap estimate $P^*((\hat{S}_n^W)^{-1}\hat{T}_n^W \leq v)$ are plotted against $P(\hat{S}_n^{-1}(\hat{T}_n - T(P)) \leq v)$. Again, this is done for values of v such that $P(\hat{S}_n^{-1}(\hat{T}_n - T(P)) \leq v) = .01$, j/20 (for j =1, ..., 19), and .99. Again, the values of v have been determined by 16 000 simulations. For the Monte Carlo estimation of expected squared error, bias , and standard deviation of bootstrap and wild bootstrap also 10 000 simulations have been used. The plots show impresssively the superiority of bootstrapping the studentized statistic over the normal approximation.

6. Proofs. We will give here only the proof of Theorem 1. The proofs of the other theorems can be carried out using the same approach, but they are much more complicated. In particular, let us shortly mention one argument used for the non i. i. d. case to show that consistency of bootstrap implies asymptotic normality. To prove here asymptotic neglibility of the observations one can proceed as follows. First one proves consistency of the bootstrap procedure with Poisson sample size (for an explicit definition of this resampling procedure see Section 4). Replacing observations by independent copies one gets now asymptotic neglibility of the observations (Use the representation (4.3) of bootstrap with Poisson sample size). Detailed proofs of Theorems 1, 2 and 3 are given in Mammen (1992).

PROOF OF THEOREM 1. The equivalence of (i) and (ii) is well known and follows from Raikov (1938). In particular, the implication "(ii) ⇒ (i)" is trivial: Note that according to (ii) there exist measures μ_n with $d_\infty(\mu_n, N(0, \hat{S}_n^2)) \to 0$ (in probability). This implies that for a sequence of positive numbers σ_n one gets $(\hat{S}_n - \sigma_n)/\sigma_n \to 0$. Therefore (i) follows. The implication "(i) ⇒ (iii)" may also be more or less known. For completeness we give also the proofs of "(i) ⇒ (ii)" and "(i) ⇒ (iii)".

In the proofs we apply the following version of the central limit theorem (see Gnedenko and Kolmogorov (1954) and Araujo and Giné (1980)) :

THEOREM. *Suppose* $U_{n,1}, \dots, U_{n,n}$ *are i. i. d. with distribution* P_n. *Then the following statements are equivalent:*

(6.1) $$\sum_{i=1}^{n} U_{n,i} \to N(0,1) \quad (\textit{in distribution})$$

(6.2) *There exists* $\tau > 0$ *with:*

(a) $$n\,E\;U_{n,i,\tau} \to 0,$$

(b) $$n\,E\,(U_{n,i,\tau} - EU_{n,i,\tau})^2 \to 1,$$

(c) $$n\,P(\,|U_{n,i}| > \varepsilon\,) \to 0 \quad \forall\, \varepsilon > 0,$$

where $U_{n,i,\tau} = U_{n,i}\ 1(\,|U_{n,i}| \le \tau\,)$.

This version of the central limit theorem says that a sum of i. i. d. observations converges to a normal limit (6.1) if and only if the maximal summand is of lower order than the sum (6.2c). Conditions (6.2a,b) are only needed to guarantee the correct expectation and variance of the approximating normal distribution.

PROOF OF "(i) ⇒ (ii)". Put $Y_{n,i} = (g_n(X_{n,i}) - t_n)/(n\sigma_n)$. Then $\hat{S}_n^2$ can be written as

$$\hat{S}_n^2 = \sigma_n^2 \sum_{i=1}^{n} \Big(Y_{n,i} - \frac{1}{n}\sum_j Y_{n,j}\Big)^2.$$

It suffices to show

(6.3) $$\hat{S}_n^2 / \sigma_n^2 \to 1 \ \text{(in probability)}.$$

Because of $\sqrt{n}\,\frac{1}{n}\sum_j Y_{n,j} = \frac{1}{\sqrt{n}}\,\frac{T_n - t_n}{\sigma_n} \to 0$ this follows from

(6.4) $$\sum_j Y_{n,j}^2 \to 1 \ \text{(in probability)}.$$

To see (6.4) we apply the central limit theorem with $U_{n,i} = Y_{n,i}$. Because of (6.2c) one can choose $\varepsilon_n \to 0$ such that $n\,P(\,|Y_{n,i}| > \varepsilon_n\,) \to 0$. Then

$$\sum_j Y_{n,j}^2 = \sum_j Y_{n,j,\,\varepsilon_n}^2 + o_P(1),$$

where $Y_{n,i,\tau} = Y_{n,i}\ 1(\,|Y_{n,i}| \le \tau\,)$ for $\tau > 0$.

Now (6.4) follows from

$$\operatorname{var}\Big(\sum_j Y_{n,j,\,\varepsilon_n}^2\Big) \le \sum_j E\,Y_{n,j,\,\varepsilon_n}^4 \le n\,E\,Y_{n,j,\varepsilon_n}^2 \cdot \varepsilon_n^2 \to 0$$

and

$$E \sum_j Y^2_{n,j,\varepsilon_n} = n \ \mathrm{var}\ (Y_{n,j,\varepsilon_n}) + n\ (E\ Y_{n,j,\varepsilon_n})^2 \to 1.$$

(see (6.2a) and (6.2b)).

PROOF OF "(i) $\Rightarrow$ (iii)". We have to show condition (6.2) of the central limit theorem for the bootstrap world. Put

$$Y^*_{n,i} = \frac{g_n\left(X^*_{n,i}\right) - \frac{1}{n}\sum_{j=1}^{n} g_n(X_{n,j})}{n\,\sigma_n}.$$

As in the proof of "(i) $\Rightarrow$ (ii)" choose $\varepsilon_n \to 0$ such that $n\ P(\mid Y_{n,i} \mid > \varepsilon_n) \to 0$. Then $P(P^*(\left|Y^*_{n,i}\right| \le 2\,\varepsilon_n) = 1) \to 1$, i. e. for most samples one gets that, conditionally given the sample, the absolutely maximal observation in the "bootstrap world" is always of lower order than the sum in the "real world". For (6.2 b) note that with probability tending to 1 for every $\tau > 0$ with $Y^*_{n,i,\tau} = Y^*_{n,i}\ 1(\mid Y^*_{n,i} \mid \le \tau)$:

$$n\,E^*\,(Y^*_{n,i,\tau} - E^*\ Y^*_{n,i,\tau})^2 = n\,E^*\,(Y^*_{n,i} - E^*\ Y^*_{n,i})^2$$

$$= \hat{S}^2_n / \sigma^2_n = 1 + o_p(1),$$

as has been shown in the proof of "(i) $\Rightarrow$ (ii)".

PROOF OF "(iii) $\Rightarrow$ (ii)". Without loss of generality suppose $g_n = \mathrm{id}$. In disagreement with old notation we put $Y_{n,i} = X_{n,i} / n$ and $Y^*_{n,i} = X^*_{n,i}/n$. We suppose

(6.5) $\qquad \mathrm{med}(P_n) \ni 0,$

(6.6) $\qquad P(\mid \hat{T}_n \mid \le 1) \ge 1/2$, and $P(\mid \hat{T}_n \mid < 1) \le 1/2$.

(6.5) and (6.6) can be achieved by replacing $X_{n,i}$ by $c_n\ X_{n,i} + b_n$ for appropriate constants c_n and b_n. The proof of (ii) is given in two steps.

STEP 1. First we show

(6.7) $$n\, P(|Y_{n,i}| > \delta) \to 0 \qquad \forall \delta > 0.$$

Proof of (6.7). In the theory of central limit theorems concentration inequalities have been proved which entail that for a sum of inedependent observations the maximal (centered) summand cannot be of larger order than the sum. We apply the following concentration inequality (see Theorem 1 on p. 407 in Le Cam (1986) and Esseen (1966)).

THEOREM (CONCENTRATION INEQUALITY): *For n independent random variables* $U_1, \dots, U_n$ *put* $R = U_1 + \dots + U_n$. *Let* μ_j *be a median of* U_j. *Then for* $\tau > 0$

(6.8) $$\sup_x P(x \le R \le x + \tau) \le \sqrt{8\pi}\, \left[\sum_{i=1}^n P(|U_i - \mu_i| \ge \tau) \right]^{-1/2}.$$

Choose $\delta > 0$. Because of (6.6) there exists a sequence x_n with

$$c = P(x_n \le \hat{T}_n \le x_n + \delta/2) > 0.$$

Then, because of (2.3),

(6.9) $$P^*(x_n \le \hat{T}_n^* - \hat{T}_n + t_n \le x_n + \delta/2) = c + o_p(1).$$

The concentration inequality (6.8) implies with $\hat{\mu} = \text{med}(Y_{n,1}, \dots, Y_{n,n})$

(6.10) $$\#\{i: |Y_{n,i} - \hat{\mu}| \ge \delta/2\} = n\, P^*(|Y_{n,i}^* - \hat{\mu}| \ge \delta/2) \le 8\pi\, c^{-2} + o_p(1).$$

(6.10) and (6.5) imply $|\hat{\mu}| \le \delta/2$ with probability tending to one. Therefore

(6.11) $$\#\{i: |Y_{n,i}| \ge \delta\} \le 8\pi\, c^{-2} + o_p(1).$$

(6.11) shows (6.7). Suppose for instance $n\, P(|Y_{n,i}| > \delta) \to \gamma$ for a δ and $\gamma > 0$. Then $\#\{i: |Y_{n,i}| \ge \delta\}$ is asymptotically Poisson distributed with parameter γ. This would contradict (6.11) for every $c > 0$.

STEP 2. We show now (2.2) $d_\infty(\mathcal{L}(\widehat{T}_n - t_n), N(0, \widehat{S}_n^2)) \to 0$ (in probability). Because of (6.7) we can find a sequence $\delta_n \to 0$ with $n P(|Y_{n,i}| > \delta_n) \to 0$. This implies $P(P^*(|Y^*_{n,i}| \le 2\delta_n) = 1) \to 1$. We apply now the central limit theorem with $U_{n,i} = (Y^*_{n,i} - \widehat{T}_n) / \widehat{S}_n$. For (2.2) we have to show that conditionally given $X_{n,1}, \ldots, X_{n,n}$ (6.2) holds with probability tending to one. This follows from:

$$\widehat{S}_n^{-1} = O_P(1). \tag{6.12}$$

It remains to show (6.12).

Proof of (6.12). Suppose that (6.12) does not hold and that $P(\widehat{S}_n < \gamma_n) > 0$ for a sequence $\gamma_n \to 0$. Because of $\widehat{S}_n^2 = \mathrm{var}^*(\widehat{T}_n^* - \widehat{T}_n)$ and (2.3) this would imply $\widehat{T}_n - t_n \to 0$ (in probability). But this would contradict (6.6).

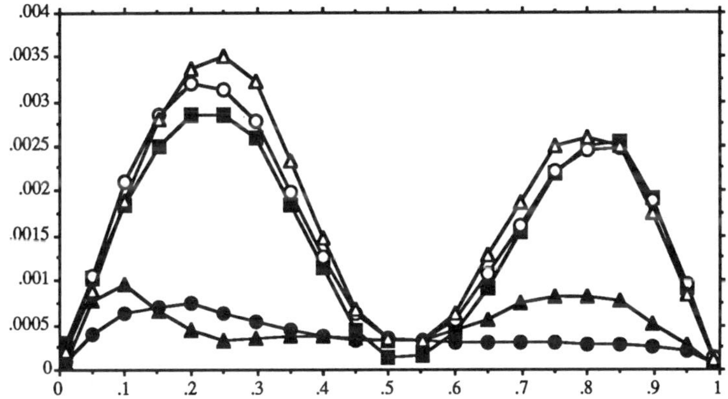

Figure 1a. *Monte Carlo estimates of the* **expected squared error** *of estimates of the distribution function of a unstudentized linear statistic and a studentized linear statistic , plotted against the distribution function.* ○ *bootstrap,* △ *wild bootstrap, and* ■ *normal approximation of the unstudentized statistic,* ● *bootstrap and* ▲ *wild bootstrap of the studentized statistic. Model A.*

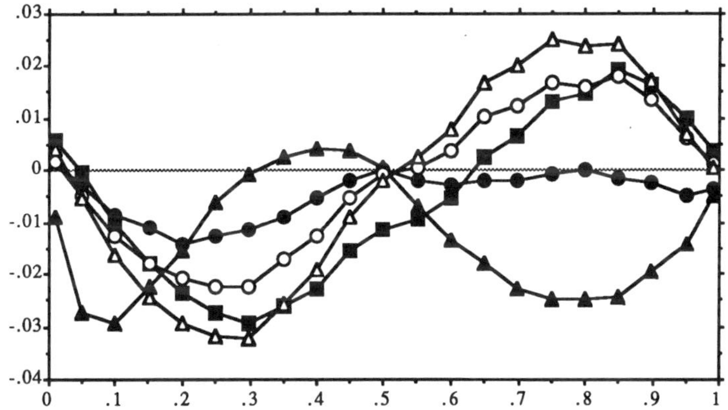

Figure 1b. *Monte Carlo estimates of the* **bias** *of estimates of the distribution function of a unstudentized linear statistic and a studentized linear statistic , plotted against the distribution function.* ○ *bootstrap,* △ *wild bootstrap, and* ■ *normal approximation of the unstudentized statistic,* ● *bootstrap and* ▲ *wild bootstrap of the studentized statistic. Model A.*

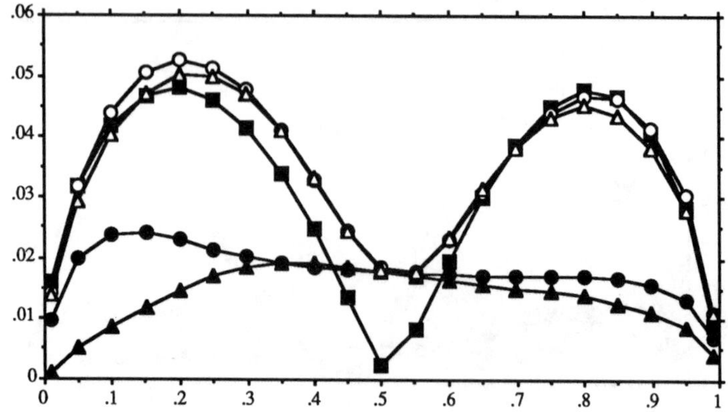

Figure 1c. *Monte Carlo estimates of the* ***standard deviation*** *of estimates of the distribution function of a unstudentized linear statistic and a studentized linear statistic , plotted against the distribution function.* ○ *bootstrap,* △ *wild bootstrap, and* ■ *normal approximation of the unstudentized statistic,* ● *bootstrap and* ▲ *wild bootstrap of the studentized statistic. Model A.*

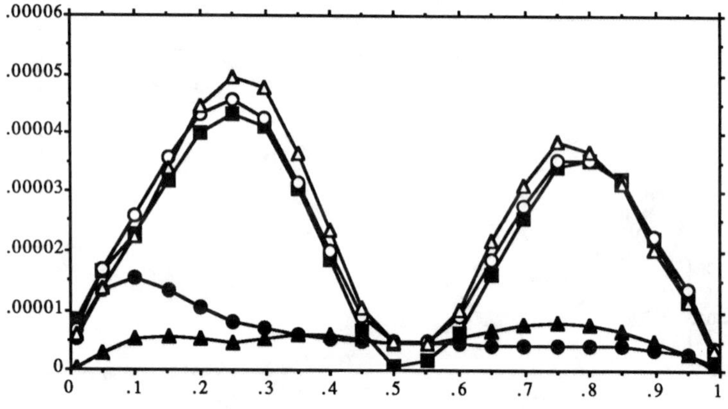

Figure 1d. *Standard deviations of the Monte Carlo calculations in Figure 1a.*

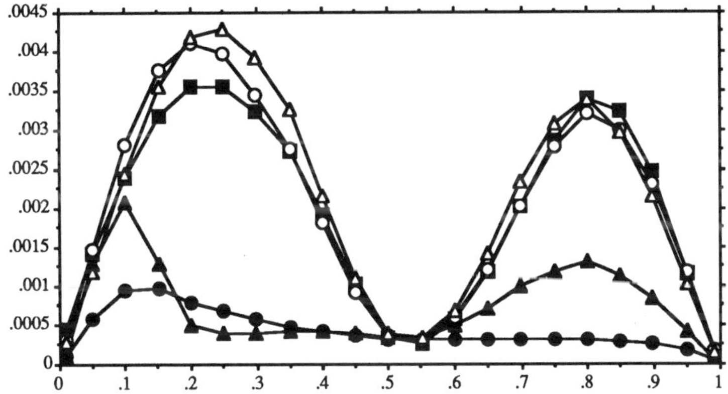

Figure 2. *Monte Carlo estimates of the* ***expected squared error*** *of estimates of the distribution function of a unstudentized linear statistic and a studentized linear statistic , plotted against the distribution function.* ○ *bootstrap,* △ *wild bootstrap, and* ■ *normal approximation of the unstudentized statistic,* ● *bootstrap and* ▲ *wild bootstrap of the studentized statistic. Model B.*

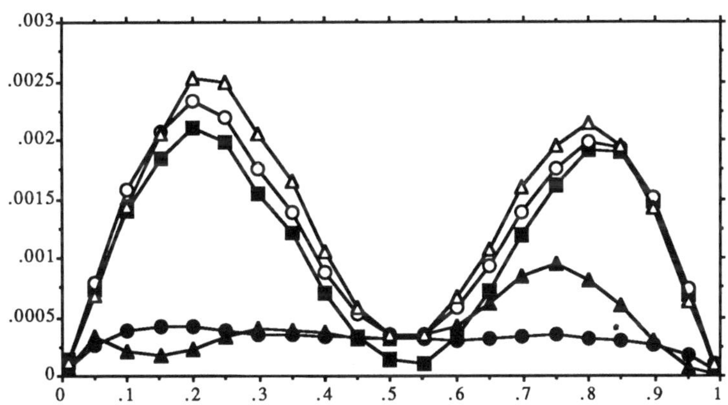

Figure 3. *Monte Carlo estimates of the* ***expected squared error*** *of estimates of the distribution function of a unstudentized linear statistic and a studentized linear statistic , plotted against the distribution function.* ○ *bootstrap,* △ *wild bootstrap, and* ■ *normal approximation of the unstudentized statistic,* ● *bootstrap and* ▲ *wild bootstrap of the studentized statistic. Model C.*

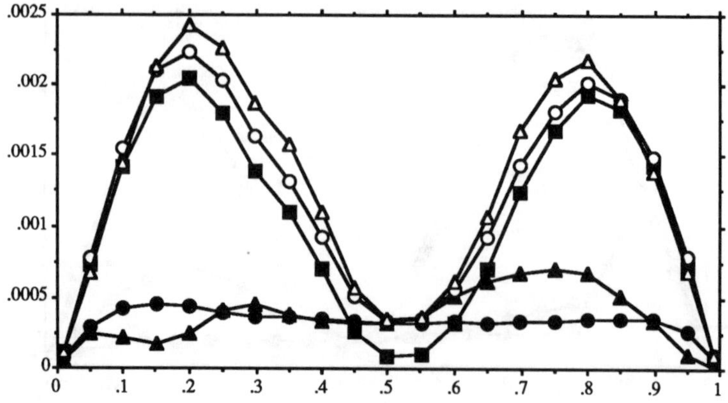

Figure 4. *Monte Carlo estimates of the* ***expected squared error*** *of estimates of the distribution function of a unstudentized linear statistic and a studentized linear statistic , plotted against the distribution function.* ○ *bootstrap,* △ *wild bootstrap, and* ■ *normal approximation of the unstudentized statistic,* ● *bootstrap and* ▲ *wild bootstrap of the studentized statistic. Model D.*

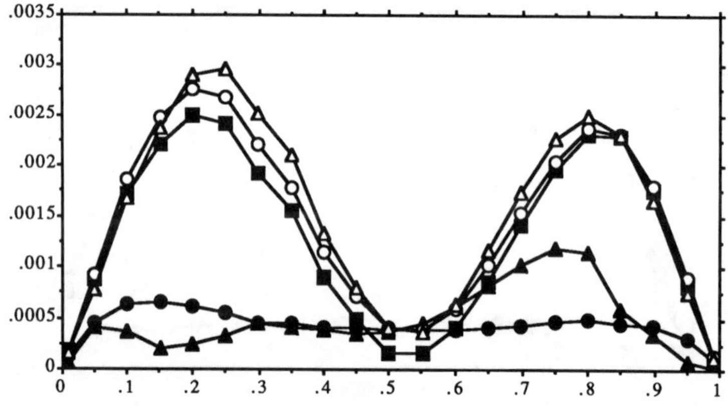

Figure 5. *Monte Carlo estimates of the* ***expected squared error*** *of estimates of the distribution function of a unstudentized linear statistic and a studentized linear statistic , plotted against the distribution function.* ○ *bootstrap,* △ *wild bootstrap, and* ■ *normal approximation of the unstudentized statistic,* ● *bootstrap and* ▲ *wild bootstrap of the studentized statistic. Model E.*

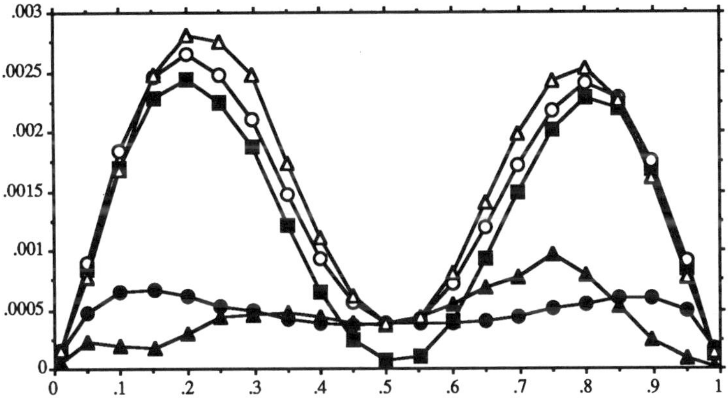

Figure 6a. *Monte Carlo estimates of the* ***expected squared error*** *of estimates of the distribution function of a unstudentized linear statistic and a studentized linear statistic , plotted against the distribution function.* ○ *bootstrap,* △ *wild bootstrap, and* ■ *normal approximation of the unstudentized statistic,* ● *bootstrap and* ▲ *wild bootstrap of the studentized statistic. Model F.*

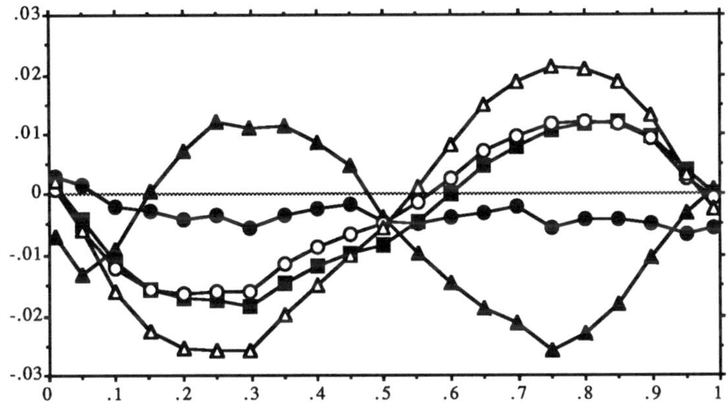

Figure 6b. *Monte Carlo estimates of the* ***bias*** *of estimates of the distribution function of a unstudentized linear statistic and a studentized linear statistic , plotted against the distribution function.* ○ *bootstrap,* △ *wild bootstrap, and* ■ *normal approximation of the unstudentized statistic,* ● *bootstrap and* ▲ *wild bootstrap of the studentized statistic. Model F.*

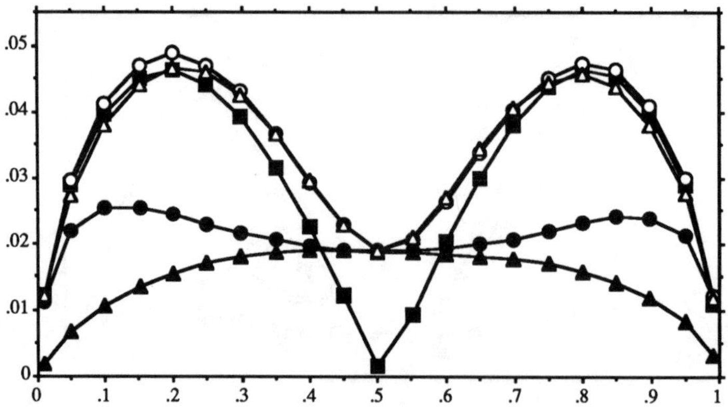

Figure 6c. *Monte Carlo estimates of the* **standard deviation** *of estimates of the distribution function of a unstudentized linear statistic and a studentized linear statistic , plotted against the distribution function.* ○ *bootstrap,* △ *wild bootstrap, and* ■ *normal approximation of the unstudentized statistic,* ● *bootstrap and* ▲ *wild bootstrap of the studentized statistic. Model F.*

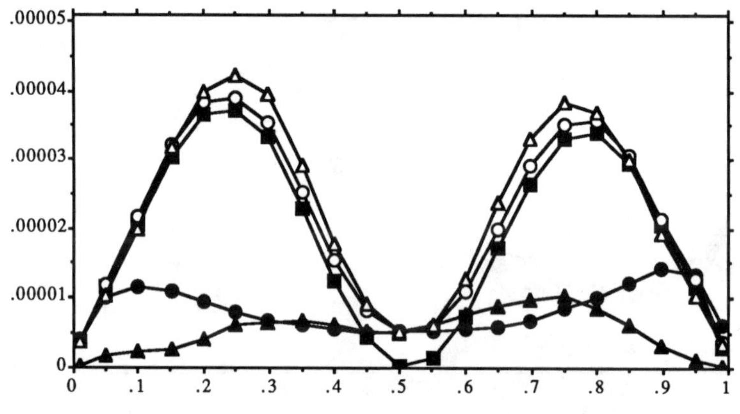

Figure 6d. *Standard deviations of the Monte Carlo calculations in Figure 6a.*

Chapter 2: An example where bootstrap fails: Comparing nonparametric versus parametric regression fits.

1. A goodness-of-fit test. In the next three chapters we discuss applications of bootstrap in nonparametric curve estimation. In this chapter we give an example of nonparametric regression where bootstrap does not work. We will show that bootstrap cannot be used for the approximate determination of critical values for a certain goodness - of - fit test statistic. The model being tested is a parametric regression model and the test statistic is based on the distance between a parametric and a nonparametric regression estimate.The material of this section is also contained in Härdle and Mammen (1990).

We consider the following model. Given are n i.i.d. observations $\{(X_i,Y_i)\}_{i=1}^n$ ($X_i \in \mathbf{R}^d$, $Y_i \in \mathbf{R}$) with unknown regression function $m(\bullet) = E(Y_i \mid X_i = \bullet)$. We write also $Y_i = m(X_i) + \varepsilon_i$ with $E(\varepsilon_i \mid X_i) = 0$. Bootstrap methods for this model have been considered in a film by McDonald (1982) and in Dikta (1988). They propose to resample from the pairs $\{(X_i,Y_i)\}_{i=1}^n$, i.e., the bootstrap sample $\{(X_i^*, Y_i^*)\}_{i=1}^n$ is drawn (with replacement) from the set $\{(X_i,Y_i)\}_{i=1}^n$. We will give an example where this bootstrap procedure does not work because it does not reflect the stochastic structure of the model: in the "bootstrap world" the conditional expectation $E^*(Y_i^* \mid X_i^*)$ is equal to Y_i^* (if the design variables X_i are pairwise different). As another resampling method we will propose wild bootstrap (see Chapter 1) which will work in our example. For another proposal which is based on pairwise resampling from a smooth distribution function see Cao - Abad and Gonzales - Manteiga (1990). Bootstrap in the model of deterministic design variables X_i and i.i.d. error variables ε_i have been considered in Härdle and Bowman (1988) and Hall (1990b, c). For an overview of applications of bootstrap in nonparametric regression see Härdle and Mammen (1991). A detailed analysis of bootstrap in nonparametric curve estimation based on higher - order Edgeworth expansions can be found in Hall (1990e).

In this chapter we want to apply bootstrap for the following testing problem. We wish to test the parametric model $\{m_\theta: \theta \in \Theta\}$ against the nonparametric alternative which only assumes that $m(\bullet)$ is "smooth". A natural approach is to plot a parametric regression estimator $m_{\hat\theta}$ against a kernel estimator $\hat m_h$ with bandwidth $h = h_n$ and kernel K (Nadaraya, 1964; Watson, 1964)

$$\widehat{m}_h(x) = \frac{\sum_{i=1}^{n} K_h(x - X_i)\, Y_i}{\sum_{i=1}^{n} K_h(x - X_i)},$$

where $K_h(\bullet) = h^{-d} K(\bullet / h)$. (To simplify the notation, the dependence of $\widehat{\theta}$ and h on n will be dropped.) For a general discussion of kernel estimates in nonparamtric regression we refer to Härdle (1990). The question is now whether the differences which may be visible between $m_{\widehat{\theta}}$ and $\widehat{m}_h$ can be explained by stochastic fluctuations within the parametric model. One way to proceed is to measure the difference between $m_{\widehat{\theta}}$ and $\widehat{m}_h$ by a distance and to use this distance as a test statistic. Here we study the L_2-distance between the nonparametric and parametric fits. The use of this distance is motivated by mathematical convenience. From a more data - analytic point of view one would prefer distances which reflect similarities in the shape of the regression functions (e.g. the number of peaks, see Mammen, 1991b) Nevertheless we restrict ourselves to the treatment of the weighted L_2-distance $\int (\widehat{m}_h - m_{\widehat{\theta}})^2 \, \pi$ where π is a weight function.

Let $\mathcal{K}_{h,n}$ denote the (random) smoothing operator

$$\mathcal{K}_{h,n}\; g(x) = \frac{\sum_{i=1}^{n} K_h(x - X_i)\, g(X_i)}{\sum_{i=1}^{n} K_h(x - X_i)}.$$

Because $E(\widehat{m}_h(x) \mid X_1, \ldots . X_n) = \mathcal{K}_{h,n}\, m(x)$ we will consider the following modification of the squared separation between $\widehat{m}_h$ and $m_{\widehat{\theta}}$:

$$\widehat{T}_n = n h^{d/2} \int \left(\widehat{m}_h(x) - \mathcal{K}_{h,n}\, m_{\widehat{\theta}}(x)\right)^2 \pi(x) dx.$$

We propose to use $\widehat{T}_n$ as a test statistic to test the parametric hypothesis:

$$m \in \{ m_\theta : \theta \in \Theta \}.$$

The idea to use nonparametric curve estimates for constructing test statistics goes back to Bickel and Rosenblatt (1973) where the L_2 - distance between a kernel estimate and its expectation under the null hypotheses has been proposed for testing a simple hypotheses. For asymptotic power considerations on this test see Ghosh and Huang (1991). Other tests using kernel estimates or other nonparametric curve estimates have been proposed in Neuhaus (1986, 1988), Azzalini, Bowman, and Härdle(1989), Cox, Koh, Wahba, and Yandell(1988), Cox and Koh(1989), Eubank and Spiegelman(1989), Firth, Glosup, and Hinkley (1989), Härdle and Marron (1990), and Raz (1990). For a more general discussion of bootstraping test statistics based on distances see also Romano (1988, 1989).

For an approximate calculation of critical values we determine the asymptotic distribution of $\hat{T}_n$ for a parametric regression function $m = m_{\theta_0}$. We make the following assumptions on the stochastic nature of the observations and the parametric estimator of the regression function.

(A1) *With probability one* X_i *lies in a compact set (w.l.o.g.* $[0,1]^d$ *). The marginal density* $f(x)$ *of* X_i *is bounded away from zero.*

(A2) $m(\cdot)$ *and* $f(\cdot)$ *are twice continuously differentiable.* π *is continously differentiable.*

(A3) $\sigma^2(x) = \text{var}\ (Y_i \mid X_i = x)$ *is bounded away from zero and from infinity and continuous.*

(A4) $E\ \exp(t\varepsilon_i)$ *is uniformly bounded in* n *for* $|t|$ *small enough (where* $\varepsilon_i = Y_i - m(X_i)$ *).*

For simplicity we consider only the k-dimensional linear parametric model. Put

$$m_\theta(x) = \theta_1 g_1(x) + \ \dots\ + \theta_k g_k(x) = < \theta,\ g(x) >$$

where g is a $\mathbf{R}^k$-valued function (for some k). The weighted least squares estimator $\hat{\theta}_n = \hat{\theta}$ is defined by

$$\hat{\theta} = \arg\ \min_\theta \sum_{i=1}^n w(X_i)(Y - m_\theta(X_i))^2$$

where w is a smooth weight function. $\hat{\theta}$ can easily be calculated

$$\hat{\theta} = \left(\sum_{i=1}^{n} w(X_i)\, g(X_i)\, g(X_i)^T\right)^{-1} \sum_{i=1}^{n} w(X_i)\, g(X_i)\, Y_i.$$

We suppose further that

(P1) w *and* g *are bounded.*

(P2) $\Sigma = E\ \ w(X_1)\ g(X_1)\ g(X_1)^T$ *is positively definite.*

For the kernel K we make the following assumptions.

(K1) *The kernel* K *is a symmetric, twice continuously differentiable probability density function with compact support.*

(K2) *The bandwidth* h *fullfills* $h = h_n \sim n^{-1/(d+4)}$.

(K2) is fulfilled in particular for every choice of the bandwidth h which is asymptotically optimal for the class of two times continuously differentiable regression functions.

In the next theorem we show that $\hat{T}_n$ is asymptotically normal on the hypothesis.

THEOREM 1. *Assume* (A1) – (A4), (P1), (P2), (K1), (K2), *and that* m *lies in the parametric hypothesis* $\{m_\theta: \theta \in \Theta\}$. *Then* $\hat{T}_n - b_h$ *converges weakly to* N(0,V), *where*

$$b_h = h^{-d/2} K^{(2)}(0) \int \frac{\sigma^2(x)\, \pi(x)}{f(x)}\, dx,$$

$$V = 2K^{(4)}(0) \int \frac{\sigma^4(x)\, \pi(x)^2}{f^2(x)}\, dx\,,$$

and where $K^{(j)}$ *denotes the* j *-fold convolution product of* K.

For asymptotic power considerations of this test we refer to Härdle and Mammen (1990).

2. How to bootstrap. Bootstrap and wild bootstrap. The proof of Theorem 1 is based on a stochastic expansion with error term of order $n^{-1/10}$. Therefore the theorem can only give a rough idea of the stochastic behaviour of $\widehat{T}_n$ if the sample size is small. Indeed, in the simulations given in Härdle and Mammen (1990) one can see that the normal approximation does not work very well for moderate sample sizes. In this section we will study bootstrap methods as an alternative to asymptotics. We consider three different possibilities of bootstrapping

- *the naive resampling method;*
- *the adjusted residuals bootstrap;*
- *the wild bootstrap.*

We show that only the third type of bootstrap works. The *naive bootstrap* consists of simply resampling the original observations: i.e., the bootstrap sample $\{(X_i^*, Y_i^*)\}_{i=1}^n$ is drawn (with replacement) from the set $\{(X_i,Y_i)\}_{i=1}^n$. We then form $\widehat{T}^{*,N}$ like $\widehat{T}_n$ as the squared separation between the parametric fit $m_{\hat{\theta}^*}$ and the non-parametric fit $\widehat{m}_h^*$ (both computed from the bootstrap sample $\{(X_i^*, Y_i^*)\}_{i=1}^n$):

$$\widehat{T}^{*,N} = n\, h^{d/2} \int \left(\widehat{m}_h^*(x) - \mathcal{K}_{h,n}\, m_{\hat{\theta}^*}(x)\right)^2 \pi(x)\, dx \ .$$

From the conditional distribution $\mathcal{L}^*(\widehat{T}^{*,N}) = \mathcal{L}(\widehat{T}^{*,N} \mid (X_i,Y_i)_{i=1}^n)$ we take the $(1-\alpha)$ quantile $\hat{t}_\alpha^N$ and reject the parametric hypothesis if $\widehat{T}_n > \hat{t}_\alpha^N$. We call this resampling procedure here *naive* because, as we will show below, it does not reflect correctly the stochastic structures of the model, at least for the purposes of this chapter.

The *adjusted residuals bootstrap* is motivated by the argument that the naive bootstrap overestimates the critical value when the parametric model is false and has therefore small power. We define the bootstrap with adjusted residuals by resampling from the observations $\{(X_i, Y_i - \widehat{m}_h(X_i) + m_{\hat{\theta}}(X_i))\}_{i=1}^n$. $\widehat{T}^{*,A}$ is now be formed like $\widehat{T}_n$ by the squared separation between the parametric fit and the non-parametric fit. As above we take $\hat{t}_\alpha^A$ as the $(1-\alpha)$ quantile of the conditional distribution $\mathcal{L}^*(\widehat{T}^{*,A}) = \mathcal{L}(\widehat{T}^{*,A} \mid \{(X_i,Y_i)\}_{i=1}^n)$ and reject the parametric hypothesis if $\widehat{T}_n > \hat{t}_\alpha^A$. In the following theorem we show that on the hypothesis $\mathcal{L}^*(\widehat{T}^{*,N})$ and $\mathcal{L}^*(\widehat{T}^{*,A})$ have different variances from $\mathcal{L}(\widehat{T}_n)$. This implies that naive bootstrap and adjusted residuals bootstrap do not work.

THEOREM 2. *Assume* (A1), ... , (A4), (K1), (K2), (P1), (P2), *and that* m *lies in the parametric hypothesis* $\{m_\theta: \theta \in \Theta\}$. *Define* $\widehat{T}^{*,N}$ *and* $\widehat{T}^{*,A}$ *as above. Then the bootstrap estimates of the variance of* T_n *converge as follows*

$$\mathrm{var}^*\left(\widehat{T}^{*,N}\right) - 3\ \mathrm{var}\left(\widehat{T}_n\right) \xrightarrow{P} 0\ ,$$

$$\mathrm{var}^*\left(\widehat{T}^{*,A}\right) - \mathrm{var}\left(\widehat{T}_n\right)$$

$$- 4 \int \frac{\sigma^4(x)\,\pi^2(x)}{f^2(x)}\,dx \int \left(K^{(2)}(t) - K^{(3)}(t)\right)^2 dt \xrightarrow{P} 0\ .$$

Theorem 2 shows that the proposed bootstrap procedures are inconsistent, even after a bias correction. At first sight this result is surprising and seems to contradict the intuitive rationale for the bootstrap. But on reflection, the reason for the failure is clear: the regression function is not equal to the conditional expectation of the observation under the bootstrap distribution, i.e. $E^* (Y_i^* - m(X_i^*) \mid X_i^*\ ) = Y_i^* - m(X_i^*) \neq 0$, in general. Here E^* denotes the conditional expectation $E(\bullet \mid \{(X_i,Y_1)\}_{i=1}^n\ )$. Wu (1986) pointed out that bootstrap can be inconsistent for least-squares estimators in linear models when the conditional variance is not constant. In our case the bootstrap breaks, even with homoscedastic errors.

For another, more mathematical explanation of the inconsistency of bootstrap note first that $\widehat{T}_n - b_h$ is asymptotically equivalent to a pure U-statistic, i.e. there exists a function H_n such that

$$\widehat{T}_n - b_h = \sum_{i \neq j} H_n((X_i,Y_i)\,,(X_j,Y_j)) + o_P(1),$$

with $E\ H_n((X_i,Y_i), (X_j,Y_j) \mid X_j,Y_j) = 0$ for $i \neq j$ (see the proof of Theorem 1). The following lemma, which we learned from van Zwet (1989), shows that bootstrap does not work for pure U-statistics as an estimate of the distribution of $V_n = \widehat{T}_n - E\widehat{T}_n$. In the next chapter we will see that bootstrap may also work for pure U-statistics after another more appropriate choice of the bootstrapped statistic V_n.

LEMMA. *For a sample* $U_1, \dots, U_n$ *of i.i.d. random variables and a symmetric function* H $(H(x,y) = H(y,x))$ *assume*

$$E(H(U_1,U_2) \mid U_2) = 0,$$

$$E\ H^2(U_1,U_1) < \infty,$$

$$E\ H^2\ (U_1,U_2) < \infty.$$

We consider

$$S\ = \Sigma_{i \neq j}\ H(U_i, U_j),$$

$$S^*\ = \Sigma_{i \neq j}\ H(U_i^*, U_j^*\ )$$

where $(U_1^*, \dots, U_n^*)$ *is a resample drawn from* $\{U_1, \dots, U_n\}$. *Then*

$$n^{-2}\ [\ E\ var^*\ (S^*) - 3\ var\ (S)\ ] \to 0.$$

Note that under the assumption of the lemma n^{-2} is the correct norming factor because of

$$var(S) = n(n-1)\ \ E\ \ H^2(U_1,U_2).$$

As an alternative resampling procedure for the test statistic T_n we propose the *wild bootstrap* (see Chapter 1). This approach attempts to mimic the conditional distribution of Y_i given X_i rather than the i. i. d. structure of (X_i,Y_i). We define

$$\tilde{\varepsilon}_i = Y_i - \hat{m}_h\,(X_i)\,.$$

This *single residual* $\tilde{\varepsilon}_i$ is used by the wild bootstrap procedure to estimate the conditional distribution $\mathcal{L}(Y_i - m\ (X_i\ ) \mid X_i)$ by an $\hat{F}_i$, say. More precisely one defines an arbitrary distribution $\hat{F}_i$ such that

$$E_{\hat{F}_i}\ Z\ =\ 0,$$

$$E_{\hat{F}_i} Z^2 = (\tilde{\varepsilon}_i)^2,$$

$$E_{\hat{F}_i} Z^3 = (\tilde{\varepsilon}_i)^3.$$

In the sequel we assume that $\hat{F}_i$ is chosen according to one of the three constructions given at the end of Chapter 1 Section 4. The following results hold also for other constructions as in (4.7), ..., (4.9) of Chapter 1 under sufficient moment conditions on the distribution Q.

Now construct independent $\varepsilon_i^W \sim \hat{F}_i$ and use $(X_i, Y_i^W = m_{\hat{\theta}}(X_i) + \varepsilon_i^W)$ as bootstrap observations. Then generate $T^{*,W}$ like $\hat{T}_n$ by the squared separation between the parametric fit and the non-parametric fit. From the Monte Carlo approximation of $\mathcal{L}^*(\hat{T}^{*,W})$ construct the $(1-\alpha)$ quantile $\hat{t}_\alpha^W$ and reject the parametric hypothesis if $\hat{T}_n > \hat{t}_\alpha^W$. In the following Theorem we show that this procedure works.

THEOREM 3. *Assume* (A1), ... , (A4), (P1), (P2), (K1), (K2), *that* m *lies in the parametric hypothesis* $\{m_\theta : \theta \in \Theta\}$. *Then* $\mathcal{L}^*\left(\hat{T}^{*,W} - b_h\right)$ *converges weakly to* N(0,V), *where* b_h *and* V *are defined in Theorem 1.*

Wild bootstrap has also been considered for the estimation of the distribution of $\hat{m}_h(x) - m(x)$ at a fixed point x (see Härdle and Marron (1990) and Cao - Abad (1991)).

Modifications of $\hat{T}^{*,W}$ are straightforward if it is appropriate to assume certain smoothness conditions for $\sigma^2(\bullet)$. Then the variance of $\hat{F}_i$ should be taken as $\hat{\sigma}^2(X_i)$ where $\hat{\sigma}^2(\bullet)$ is a nonparametric estimator of $\sigma^2(\bullet)$, see Carroll (1982).

3. Proofs. W.l.o.G. we will give the proofs only for d = 1 and $\pi(x) \equiv 1$.

PROOF OF THEOREM 1. First note that

$$\hat{m}_h(x) = \frac{n^{-1} \sum_{i=1}^n K_h(X_i - x)\, Y_i}{\hat{f}_h(x)},$$

where

$$\hat{f}_h(x) = n^{-1} \Sigma_{i=1}^{n} K_h(X_i - x)$$

is the kernel density estimate of the density f of the the design variables X_i. For d = 1 it is well known that

$$\hat{f}_h(x) = f(x) + O_p(n^{-2/5} \sqrt{\log n}) \text{ (uniformly in x)},$$

see Collomb and Härdle (1986),

$$\hat{m}_h(x) = m(x) + O_p(n^{-2/5} \sqrt{\log n}) \text{ (uniformly in x)},$$

see Mack and Silverman (1982). Let us briefly also consider the case d > 1. First we show for q > 0

$$P(n^{-1} \Sigma_{i=1}^{n} K_h(X_i - x) (Y_i - m(X_i)) \geq C\, n^{-2/(4+d)} \sqrt{\log n}) = O(n^{-q})$$

if C is large enough. This follows by the following simple application of the Markov inequality. Choose k > 2 + d/2. Then with $A_n = n^{2/(4+d)-1} \sqrt{\log n}\; K_h(X_1 - x)\, \varepsilon_1$ and with a constant c it holds that

$$P(n^{-1} \Sigma_{i=1}^{n} K_h(X_i - x) (Y_i - m(X_i)) \geq C\, n^{-2/(4+d)} \sqrt{\log n})$$

$$\leq \{ E \exp([n^{2/(4+d)} \sqrt{\log n}] [n^{-1} K_h(X_1 - x)\, \varepsilon_1]) \}^n \exp(-C \log n)$$

$$\leq \{ 1 + E[A_n] + \ldots + 1/k!\, E[A_n]^k [1 + \exp(A_n)]\}^n\, n^{-C}$$

$$\leq \{ 1 + c\ (\log n) / n + o((\log n) / n)\}^n\, n^{-C}$$

$$\leq n^{c - C + o(1)}.$$

For C large enough this is of order $O(n^{-q})$.

Similarly one can treat $n^{-1} \Sigma_{i=1}^{n} K_h(X_i - x)\, (m(X_i) - Y_i)$, $\hat{f}_h(x) - f(x)$, and $f(x) - \hat{f}_h(x)$. Now note that $\partial/\partial x\, \hat{f}_h(x)$, $\partial/\partial x\, f(x)$, and $\partial/\partial x\, [n^{-1} \Sigma_{i=1}^{n} K_h(X_i - x)\, m(X_i)]$ are bounded by deterministic constants with polynomial growth and that

$$\partial/\partial x\, [n^{-1} \Sigma_{i=1}^{n}\ K_h(X_i - x)\, Y_i] = O_P(h^{-2} \log n)$$

because of (A4). But this implies

$$\hat{f}_h(x) = f(x) + O_p(n^{-2/(4+d)} \sqrt{\log n}\ )\ \text{(uniformly in x)},$$

and

$$\hat{m}_h(x) = m(x) + O_p(n^{-2/(4+d)} \sqrt{\log n}\,)\ \text{(uniformly in x)}.$$

These uniform rates give

$$\hat{T}_n = n\sqrt{h} \int_0^1 (\hat{m}_h(x) - \mathcal{K}_{h,n}\, m_{\hat{\theta}}(x))^2\, dx$$

$$= n\sqrt{h} \int_0^1 (\hat{m}_h(x) - \mathcal{K}_{h,n}\, m_{\hat{\theta}}(x))^2 \left(\frac{\hat{f}_h(x)}{f(x)}\right)^2 dx\ +\ o_p(1)$$

$$= n\sqrt{h} \int_0^1 \frac{\left(\frac{1}{n} \Sigma_{i=1}^{n}\ K_h(X_i - x)\left(m(X_i) + \varepsilon_i - m_{\hat{\theta}}(X_i)\right)\right)^2}{f^2(x)}\, dx + o_p(1).$$

Now apply (P1) and (P2). Then one gets

$$\hat{T}_n = n\sqrt{h} \int_0^1 (U_{n,1}(x) + U_{n,2}(x))^2\, dx + o_p(1)$$

where

$$U_{n,1}(x) = \frac{\frac{1}{n} \Sigma_{i=1}^{n} K_h(X_i - x)\, \varepsilon_i}{f(x)},$$

$$U_{n,2}(x) = \frac{-\frac{1}{n}\sum_{i=1}^n K_h(X_i - x)\ \frac{1}{n}\sum_{j=1}^n g(X_i)^T \Sigma^{-1} w(X_j)\, g(X_j)}{f(x)}.$$

By straightforward calculations one gets

$$E\ n\sqrt{h} \int_0^1 U_{n,2}^2(x)\,dx \to 0,$$

$$E\ n^2\ h\left[\int_0^1 U_{n,1}(x)\, U_{n,2}(x)\,dx\right]^2 \to 0.$$

This implies:

$$\widehat{T}_n = n\ \sqrt{h} \int_0^1 U_{n,1}^2(x)\ dx + o_p(1)$$

$$= \widehat{T}_{n,1} + \widehat{T}_{n,2} + o_p(1),$$

where

$$\widehat{T}_{n,1} = \frac{\sqrt{h}}{n} \int_0^1 \frac{\sum_{i=1}^n K_h(X_i - x)^2\, \varepsilon_i^2}{f^2(x)}\ dx,$$

$$\widehat{T}_{n,2} = \frac{\sqrt{h}}{n} \int_0^1 \frac{\sum_{i \neq j} K_h(X_i - x)\, K_h(X_j - x)\, \varepsilon_i\, \varepsilon_j}{f^2(x)}\ dx.$$

We will show

(3.1) $$\widehat{T}_{n,1} = b_h + o_p(1),$$

(3.2) $$\mathcal{L}(\widehat{T}_{n,2}) \Rightarrow N(0,V) \quad \text{(weakly)}.$$

But (3.1), (3.2) entail the statement of Theorem 1. It remains to show (3.1) and (3.2).

Proof of (3.1). First note, that

$$E\,\hat{T}_{n,1} = E\,\sqrt{h}\int_0^1 \frac{K_h(X_1 - x)^2}{f^2(x)}\,\sigma^2(x)\,dx$$

$$= \int_0^1\int_0^1 \sqrt{h}\,\frac{K_h(u - x)^2}{f^2(x)}\,f(u)\,\sigma^2(x)\,dx$$

$$= b_h + o(1)\,.$$

Because of $\mathrm{var}(\hat{T}_{n,1} \mid X_1, \dots, X_n) = O_P((hn)^{-1}) = o_P(1)$ this implies (3.1).

Proof of (3.2). Put

$$W_{ijn} = \begin{cases} \dfrac{\sqrt{h}}{n}\displaystyle\int_0^1 \frac{K_h(X_i-x)\;K_h(X_j-x)}{f^2(x)}\,dx\;\varepsilon_i\,\varepsilon_j\,, & \text{if } i \neq j \\[2ex] 0\,, & \text{otherwise}\,. \end{cases}$$

Then

$$\hat{T}_{n,2} = \Sigma_{i,j}\;W_{ijn}\;.$$

According to Theorem 2.1 in de Jong (1987) for (3.2) it suffices to prove:

(3.3) $$\mathrm{var}\,(\hat{T}_{n,2}) \to V\,,$$

(3.4) $$\max_{1\le i\le n}\;\Sigma_{j=1}^{n}\;\mathrm{var}\,(W_{ijn}) / \mathrm{var}\,(\hat{T}_{n,2}) \to 0\,,$$

(3.5) $$E\,\hat{T}_{n,2}^4 / \left(\mathrm{var}\left(\hat{T}_{n,2}\right)\right)^2 \to 3\,.$$

The proof of (3.3) is straightforward. (3.4) follows from

$$\operatorname{var}(W_{ijn}) = O(\,n^{-2}\,)\,.$$

For the proof of (3.5) note that

$$E\ \widehat{T}^4_{n,2} = 12\ \ \Sigma^{\neq}\ EW^2_{ijn}\ W^2_{kl\,n}$$

$$+\,8\,\Sigma^{\neq}\ W^4_{ijn} + 48\ \Sigma^{\neq}_{i,j,k,l}\ E\ W_{ijn}\ W_{jkn}\ W_{kln}\ W_{lin}$$

$$+\,192\ \Sigma^{\neq}\ \ E\ W_{ijn}\ W^2_{ikn}\ W_{jkn}$$

$$= 3\operatorname{var}\left(\left(\widehat{T}_{n,2}\right)\right)^2 + o(1),$$

(Here $\Sigma^{\neq}$ denotes summation over all pairwise different indices) because of

$$E\,W^4_{12n} = O\!\left(\frac{1}{n^4\,h^2}\right) = o\!\left(\frac{1}{n^2}\right),$$

$$E\ W_{12n}\ W_{23n}\ W_{34n}\ W_{41n}$$

$$= \frac{h^2}{n^4}\int \frac{K_h(u_1 - x_1)\,K_h(u_2 - x_1)\,K_h(u_2 - x_2)\ldots K_h((u_1 - x_4))}{f^2(x_1)\ldots f^2(x_4)}$$

$$\cdot\ \ f(u_1)\ldots f(u_4)\,dx_1\ \ldots\ dx_4\ du_1\ \ldots\ du_4\cdot\ \ O(1)$$

$$= O\left(\frac{h^2}{n^4}\right)\int K^2_h\,(u_1 - u_2)\,K^{(2)}_h(u_2 - u_3)\,K^{(2)}_h(u_3 - u_4)\,K^{(2)}_h(u_4 - u_1)\,du_1\ \ldots\ du_4$$

$$= O\left(\frac{h^2}{n^4}\right)\int \left(K^{(4)}_h(u)\right)^2\ du = O\!\left(\frac{h}{n^4}\right) = o\!\left(\frac{1}{n^4}\right),$$

$$E\ W_{12n}\ W^2_{23n}\ W_{31\,n} = O\!\left(\frac{h^2}{n^4}\right)\int K_h(u^2)^2\,K^{(2)}_h(u)du$$

$$= O\!\left(\frac{1}{n^4}\right) = o\!\left(\frac{1}{n^3}\right),$$

and

$$\operatorname{var}\left(\widehat{T}_{n,2}\right) = 2 \ \Sigma^{\neq} \ E \, W_{ijn}^2 .$$

We only sketch the proofs of Theorem 2 and of Theorem 3.

PROOF OF THEOREM 2. First by lengthy but straightforward calculations using arguments of the proof of Theorem 1 one can show for $\widehat{T}^* = \widehat{T}^{*,N}$ (or $\widehat{T}^* = \widehat{T}^{*,A}$) that

$$\widehat{T}^* = \frac{\sqrt{h}}{n} \sum_{1 \le i,j \le n} \frac{K_h^{(2)}\left(X_i^* - X_j^*\right) \eta_i^* \eta_j^*}{f\left(X_i^*\right) f\left(X_j^*\right)} + \Gamma$$

for a random variable Γ with

$$E^* \, \Gamma^2 = o_p(1) .$$

Here E^* denotes the conditional expectation given $\{(X_i, Y_i)\}_{i=1}^n$. $\{(X_i^*, \eta_i^*)\}_{i=1}^n$ are drawn (with replacement) from the set $\{(X_i, \eta_i)\}$ where

$$\eta_i = \varepsilon_i + (m - m_{\hat{\theta}})(X_i) \qquad \text{if } T^* = T^{*,N} ,$$

$$\eta_i = \varepsilon_i + (m - \mathcal{K}_{h,n} \, m)(X_i) - \frac{1}{n} \sum_{j=1}^n \frac{K_h(X_i - X_j) \, \varepsilon_j}{\hat{f}_h(X_i)} \qquad \text{if } \widehat{T}^* = \widehat{T}^{*,A} .$$

Define

$$A_{ij} = \frac{K_h\left(X_i^* - X_j^*\right) \eta_i^* \eta_j^*}{f\left(X_i^*\right) f\left(X_j^*\right)} .$$

Put $a = E^*(A_{ij})$ for $i \neq j$. Then one gets with $a = O_p(n^{-1} h^{-1})$

$$\operatorname{var}^*\left(\widehat{T}^*\right) = \frac{h}{n^2} \operatorname{var}^* \left(\sum_{i \neq j} A_{ij}\right) + o_p(1)$$

$$= \frac{h}{n^2} \operatorname{var}^* \left(\sum_{i \neq j} A_{ij} \right) + o_p(1)$$

$$= \frac{h}{n^2} E^* \left(\sum_{i \neq j} A_{ij} - a \right)^2 + o_p(1)$$

$$= \frac{h}{n^2} E^* \, 2 \sum_{i \neq j} (A_{ij} - a)^2$$

$$+ \frac{h}{n^2} E^* \, 4 \sum_{i \neq j \neq k \neq i} (A_{ij} - a)(A_{jk} - a) + o_p(1)$$

$$= 2\, h\, E^* \, A_{12}^2 + 4 h n E^* \, A_{12} A_{23} + o_p(1)$$

$$= 2 \frac{h}{n^2} \sum_{i,j} \frac{K_h(X_i - X_j)^2}{f^2(X_i) f^2(X_j)} \eta_i^2 \eta_j^2$$

$$+ 4 \frac{h}{n^2} \sum_{i,j,k} \frac{K_h(X_i - X_j) K_h(X_j - X_k)}{f(X_i) f^2(X_j) f(X_k)} \eta_i \eta_j^2 \eta_k + o_p(1)$$

This gives for $\hat{T}^* = \hat{T}^{*,N}$ by straightforward calculations

$$\operatorname{var}^*\left(\hat{T}^{*,N}\right) = 6 \frac{h}{n^2} \sum_{i,j} \frac{K_h(X_i - X_j)^2}{f^2(X_i) f^2(X_j)} \varepsilon_i^2 \varepsilon_j^2 + o_p(1).$$

But the first term of the right side converges in probability to its expectation. This proves the first statement of Theorem 2. The second statement follows by a very lengthy evaluation of the above approximation of $\operatorname{var}^*\left(\hat{T}^{*,A}\right)$.

PROOF OF THEOREM 3. The proof goes along the lines of the proof of Theorem 1. In particular (A4) entails $\sup_i \varepsilon_i^2 = O_p(\log n)$ and $E \mid \varepsilon_i \mid^8 <$ const. (uniformly in n). This can be used to prove the two conditions of the above mentioned theorem of de Jong(1987).

3. A bootstrap success story: Using nonparametric density estimates in k - sample problems.

1. Bootstrap tests. In this and in the next chapter we will study applications of bootstrap in nonparametric density estimation. For other applications of bootstrap in nonparametric density estimation see Hall (1990e). We consider the following setup. Given are k data sets of i. i. d. observations

$$X_{1,1}, \dots, X_{1,n_1} \text{ with density } f_1,$$

$$\vdots \qquad \vdots$$

$$X_{k,1}, \dots, X_{k,n_k} \text{ with density } f_k.$$

We imagine the situation of a data analysis of these data sets based on the inspection of the graphical plots of k kernel density estimates

$$\hat{f}_{j,h(j)}(x) = n_j^{-1} \sum_{i=1}^{n_j} K_{h(j)}(X_{j,i} - x)$$

($j = 1, \dots, k$) where K_h is a kernel function with bandwidth h, i.e. $K_h(u) = h^{-1} K(u/h)$, and

(A1) K *is a symmetric twice continuously differentiable probability density with compact support (say* $[-1,1]$ *).*

In this context the question may arise if the data have a common density $f_1 = \dots = f_k$, with the differences between the estimated density plots being explained by random fluctuations. In this situation - i.e. after the inspection of the density estimates - a test statistic for the null hypothesis $f_1 = \dots = f_k$ should be based on the density estimates. When comparing densities, we recommend using the same amount of smoothing (i.e. the same bandwidth $h = h(j)$) in all samples $1 \le j \le k$. (For instance one might choose an average of the bandwidths chosen at first.). For a comparison of the i-th and the j-th data set we consider the L_2-distance between the corresponding kernel estimates

$$T_{ij}(h) = \int \left(\hat{f}_{i,h}(x) - \hat{f}_{j,h}(x)\right)^2 dx.$$

For references to related test statistics see the last chapter.

Note that $T_{ij}(h)$ can easily be calculated and that this involves no calculation of an integral because of

$$(1.1) \qquad T_{ij}(h) = \int \left[n_i^{-1} \sum_{u=1}^{n_i} K_h(X_{i,u} - x) \; - \; n_j^{-1} \sum_{v=1}^{n_j} K_h(X_{j,v} - x) \right]^2 dx$$

$$= n_i^{-2} \sum_{u,v=1}^{n_i} K_h^{(2)}(X_{i,u} - X_{i,v})$$

$$- 2\, n_i^{-1}\, n_j^{-1} \sum_{u=1}^{n_i} \sum_{v=1}^{n_j} K_h^{(2)}(X_{i,u} - X_{j,v})$$

$$+ \; n_j^{-2} \sum_{u,v=1}^{n_j} K_h^{(2)}(X_{j,u} - X_{j,v})$$

where $K_h^{(2)}$ is the convolution kernel $K_h^{(2)}(x) = (K_h * K_h)(x) = h^{-1}(K * K)(x/h)$.

We will consider a studentized version of $T_{ij}(h)$. Because on the hypothesis $f_i = f_j$ the diagonal term $(n_i^{-1} + n_j^{-1})\, K_h^{(2)}(0)$ is of larger order than the stochastic term $T_{ij}(h) - ET_{ij}(h)$ we substract this term from $T_{ij}(h)$ before we studentize $T_{ij}(h)$. For $f_i = f_j$ the variance of $T_{ij}(h)$ is asymptotically equivalent to $2(n_i^{-1} + n_j^{-1})^2\, h^{-1}\, K^{(4)}(0) \int f^2(x)\, dx$, where $K^{(4)}$ denotes the four - fold convolution kernel $K^{(4)} = K*K*K*K$. We will consider the following bias - corrected and studentized version of $T_{ij}(h)$

$$\Delta_{ij}(h) = [T_{ij}(h) - (n_i^{-1} + n_j^{-1})\, K_h^{(2)}(0)]$$

$$(n_i^{-1} + n_j^{-1})^{-1}\, [2K^{(4)}(0)]^{-1/2}\, h^{1/2}\, [\int (\hat{f}_h(x))^2 \; dx - n^{-1}\, K_h^{(2)}(0)\;]^{-1/2}$$

where

$$n = n_1 + \ldots + n_k$$

and

$$\hat{f}_h(x) = n^{-1} \sum_{u=1}^{k} n_u\, \hat{f}_{u,h}(x)$$

is the kernel density estimate based on the pooled sample $(X_{1,1}, \ldots, X_{k,n_k})$.

The next theorem describes the asymptotic distribution of $\Delta_{ij}(h)$ on the null hypothesis for $h = h_n \to 0$ (We will use $n = n_1 + \ldots + n_k$ also as running index.).

THEOREM 1. *Suppose* $h = h_n \to 0$, $nh_n^2 \to \infty$ *and*

(A2) $f = f_1 = \ldots = f_k$ *has compact support (say* $[0,1]$*) and is bounded away from* 0 *and* ∞.

(A3) *Assume that* $n = n_1 + \ldots + n_k \to \infty$ *and that* $n_j / n \to c_j$ *for a* c_j *in* $(0,1)$.

Then the vector $(\Delta_{ij}(h): i < j)$ *converges weakly to a mean - zero normal distribution with (possibly degenerate) covariance matrix* $\sum_{i,j;u,v}$, *where*

$$\Sigma_{i,j;u,v} = \begin{cases} 1, & \text{if } i = j \text{ and } u=v, \\ c_w^{-2} (c_i^{-1} + c_j^{-1})^{-1} (c_u^{-1} + c_v^{-1})^{-1}, & \text{if } \{i,j\} \text{ and } \{u,v\} \text{ have one common element } w, \\ 0 & \text{otherwise.} \end{cases}$$

The situation in Theorem 1 is not realistic because the bandwidth of the kernel density estimate is usually random and depends on the observations. For the case of a random data - adaptive bandwidth $\hat{h}_n = \hat{h}_n(X_{1,1}, \ldots, X_{k,n_k})$ we show in the next theorem that the asymptotic normality result of Theorem 1 can still be applied if the random bandwidth $\hat{h}_n$ stabilizes asymptotically in the following sense.

(A4) *There exists a (deterministic) sequence* $h_n \to 0$ *with* $n\, h_n^2 \to \infty$ *and* $(\hat{h}_n - h_n) / h_n = o_p(1)$.

THEOREM 2. *Assume* (A1), ... , (A4). *Then*

$$\max_{1 \le i,j \le k} \left| \Delta_{ij}(\hat{h}_n) - \Delta_{ij}(h_n) \right| \to 0 \qquad \textit{(in probability)}.$$

We propose to use $\Delta(\hat{h}_n)$ (or $\Delta(h_n)$) as a test statistic for the hypothesis

$f_1 = \dots = f_k$ where $\Delta(h) = \max_{i<j} \Delta_{ij}(h)$. Asymptotic critical values for $\Delta(h)$ can be determined with the help of Theorem 1 by simulation of normal random variables (X_{ij} : $i < j$) with mean 0 and covariance matrix $\Sigma_{i,j;u,v}$. A more direct approach which seems more attractive and more appropriate is the permutation test based on $\Delta(h)$. Or one may approximate the critical values of the test statistic by bootstrap estimates. For this purpose draw bootstrap resamples ($X^*_{1,1}, \dots , X^*_{k,n_k}$) from the set $\{X_{1,1}, \dots , X_{k,n_k}\}$ and calculate kernel density estimates based on the bootstrap samples

$$\hat{f}^*_{j,h}(x) = n_j^{-1} \sum_{i=1}^{n_j} K_h(X^*_{j,i} - x).$$

The bootstrap test statistic is defined as

$$\Delta^*(h) = \max_{i<j} \Delta^*_{ij}(h)$$

where

$$\Delta^*_{ij}(h) = [T^*_{ij}(h) - \left(n_i^{-1} + n_j^{-1}\right) K_h^{(2)}(0)]$$

$$\left(n_i^{-1} + n_j^{-1}\right)^{-1} [2K^{(4)}(0)]^{-1/2}\, h^{1/2} \left[\int \left(\hat{f}_h^*(x)\right)^2 dx - n^{-1} K_h^{(2)}(0) \right]^{-1/2},$$

$$T_{ij}^*(h) = \int \left(\hat{f}^*_{i,h}(x) - \hat{f}^*_{j,h}(x)\right)^2 dx ,$$

$$\hat{f}_h^*(x) = n^{-1} \sum_{u=1}^{k} n_u \hat{f}^*_{u,h}(x) .$$

The next theorem states that bootstrap works (asymptotically).

THEOREM 3. *Assume* (A1), (A2), (A3). *Suppose* $h_n \to 0$ *and* $n h_n^2 \to \infty$. *Then for* $1 \le i < j \le k$:

$$d_L(\mathcal{L}(\Delta_{ij}(h_n)),\ \mathcal{L}^*(\Delta^*_{ij}(h_n)) \to 0 \qquad \textit{(in probability)},$$

$$d_L(\mathcal{L}(\Delta(h_n)),\ \mathcal{L}^*(\Delta^*(h_n)) \to 0 \qquad \textit{(in probability)}.$$

Here d_L *is the Levy distance*

$d_L(\mu,\nu) =$

$$\inf\{\varepsilon: \mu(X \le t) \le \nu(X \le t+\varepsilon)+\varepsilon, \nu(X \le t) \le \mu(X \le t+\varepsilon)+\varepsilon \text{ for all } t\}.$$

In Theorem 3, and also in Theorems 4 and 6 below (but not in Theorem 5), any distance, which metricises the weak topology can be used instead of d_L. Note that $\Delta_{ij}(h_n)$ is asymptotically normal.

In the case of a random bandwidth $\hat{h}_n = \hat{h}_n(X_{1,1}, \dots, X_{k,n_k})$ one deduces immediately from Theorem 2 and Theorem 3:

THEOREM 4. *Assume* (A1) – (A4). *Then for* $1 \le i < j \le k$

$$d_L(\mathcal{L}(\Delta_{ij}(\hat{h}_n)), \mathcal{L}^*(\Delta_{ij}^*(\hat{h}_n))) \to 0 \qquad \textit{(in probability)},$$

$$d_L(\mathcal{L}(\Delta(\hat{h}_n)), \mathcal{L}^*(\Delta^*(\hat{h}_n))) \to 0 \qquad \textit{(in probability)}.$$

Note that we use the bandwidth $\hat{h}_n = \hat{h}_n(X_{1,1}, \dots, X_{k,n_k})$ and *not* $\hat{h}_n^* = \hat{h}_n(X_{1,1}^*, \dots, X_{k,n_k}^*)$ in the bootstrap resampling. This has been done for two reasons. First it would need a lot of computer calculations if in every bootstrap loop the bandwidth had to be calculated again. Secondly, especially for moderate sample sizes, it is not clear whether the whole stochastic structure of the real world should be mimicked in the bootstrap world when this structure is complicated. In practice simple models often give better results than more accurate, but more complicated, alternatives. Another problem of this type will be discussed in Chapter 8. There in the simulation of a high - dimensional linear random design model bootstrap of the studentized least squares estimator produces no more accurate results than bootstrap of the unstudentized estimator. This simulation result contradicts higher - order expansions which predict the superiority of bootstrap of the studentized estimator. In this example bootstrap seems not to mimic satisfactorily the stochastic behaviour of the least squares estimate *and* the variance estimate.

Nevertheless, for *very large* sample sizes, we expect that recalculating the bandwidth $\hat{h}_n^*$ in every bootstrap loop should provide more accurate results. Indeed below we will see that in the case of using $\hat{h}_n^*$ bootstrap works (*asymptotically*) under weaker conditions.

The proof of Theorem 3 uses the strong approximation based on the quantile transformation. Let $\widehat{F}_n$ be the empirical distribution function of the pooled sample $(X_{1,1}, \dots, X_{k,n_k})$. Let F denote the distribution with density $f = f_1 = \dots = f_k$. Consider a sample of n conditionally (given $(X_{1,1},\dots,X_{k,n_k})$) independent random variables $Y_1, \dots, Y_n$ with conditional distribution $\mathcal{L}^*(Y_1) = F$. Put $Y_i^* = \widehat{F}_n^{-1} F(Y_i)$. Then $\mathcal{L}^*(Y_i^*) = \widehat{F}_n$. The proof of Theorem 3 will be based on simple calculations estimating the distance between the statistic Δ_{ij} calculated with the sample $(Y_1, \dots, Y_n)$ and calculated with the sample $(Y_i^*, \dots, Y_n^*)$. Note that in this approach it is unnecesary to compute the asymptotic distribution of Δ_{ij}.

Let us come back to the discussion of whether the randomness of $\widehat{h}_n$ should be mimicked in the bootstrap resampling (i.e., if $\widehat{h}_n^*$ should be redetermined in every bootstrap loop). Let $\widetilde{h}_n = \widehat{h}_n(Y_1, \dots, Y_n)$ and $\widetilde{h}_n^* = \widehat{h}_n(Y_i^*, \dots, Y_n^*)$. In the next theorem we will show that bootstrap works under the following condition if one uses the bandwidth $\widehat{h}_n^*$ instead of $\widehat{h}_n$ in the bootstrap resampling.

(A4') *There exists a (deterministic) sequence* h_n *with* $h_n \to 0$ *and* $nh_n^2 \to \infty$ *such that*

$$(\widetilde{h}_n^* - \widetilde{h}_n) / h_n = o_p(1),$$

$$(\widetilde{h}_n - h_n) / h_n = O_p(1),$$

$$[(\widetilde{h}_n - h_n) / h_n]^{-1} = O_p(1).$$

Condition (A4') states that the difference between $\widetilde{h}_n$ and $\widetilde{h}_n^*$ is asymptotically neglible. It does not imply that the bandwidth $\widehat{h}_n$ stabilizes in the sense of (A4). We have included here the case that the fluctuation of $\widehat{h}_n$ are allowed to be of the same order as $\widehat{h}_n$ itself because for many practically used bandwidth selector rules, $\widehat{h}_n$, the rate of convergence of $(\widehat{h}_n - h_n) / h_n$ is only of the order $n^{-1/10}$ (see Hall and Marron, 1987, 1991). This convergence to 0 may not be visible for moderate sample sizes. Recently some proposals have been made for bandwidth selectors with fluctuations of smaller order (see Hall, Sheather, Jones and Marron (1991), Hall, Marron and Park (1989), Park and Marron (1990)).

THEOREM 5. *Assume* (A1) – (A3), (A4'). *Then for* $1 \le i < j \le k$

$$d_L(\mathcal{L}(\Delta_{ij}(\widehat{h}_n)), \mathcal{L}^*(\Delta^*_{ij}(\widehat{h}^*_n))) \to 0 \qquad \textit{(in probability)},$$

$$d_L(\mathcal{L}(\Delta(\widehat{h}_n)), \mathcal{L}^*(\Delta^*(\widehat{h}^*_n))) \to 0 \qquad \textit{(in probability)}.$$

Note that under the conditions of Theorem 5 the asymptotic result of Theorem 1 need not hold for $\Delta_{ij}(\widehat{h}_n)$ because the bandwidth $\widehat{h}_n$ may not stabilize in the sense of (A4).

In the next section we consider bootstrap confidence regions. These regions can be used for the determination of the differences between samples which are significant.

2. Bootstrap confidence regions. Tests alone are insufficient for comparing several data sets. First of all, P - values give no information about the reasons for rejection of the null hypothesis: the differences detected between the underlying densities may be trivial from a practical point of view, or they may be vital. Secondly the tests considered in the last section do not answer which differences between samples are significant. As an alternative we consider in this section bootstrap confidence regions for the set of differences $(f_i - f_j\colon\ i < j)$. The confidence regions can be used to reject all hypothesis $f_i = f_j\ (i < j)$ when the function $f \equiv 0$ is contained in none of these confidence regions.

For the construction of the confidence regions we consider the following asymptotic pivotal quantity

$$\pi(h) = \max_{i<j} \pi_{ij}(h)$$

where

$$\pi_{ij}(h) = \Big[\int \{\hat{f}_{i,h}(x) - \hat{f}_{j,h}(x) - \mathcal{K}_h(f_i - f_j)\,(x)\,\}^2\, dx - \left(n_i^{-1} + n_j^{-1}\right) K_h^{(2)}(0)\,\Big]$$

$$\left[\, 2K^{(4)}(0) \cdot h^{-1} \int \left(n_i^{-1}\, \hat{f}_{i,h}(x) + n_j^{-1}\, \hat{f}_{j,h}(x)\right)^2 dx \,\right]^{-1/2},$$

and where $\mathcal{K}_h$ denotes the following smoothing operator

$$\mathcal{K}_h \; g(x) = \int K_h(x - u) \, g(u) \, du \, .$$

Note that compared with $\Delta_{ij}(h)$ we have now a slightly modified factor $[\ldots]^{-1/2}$ because we do not assume $f_i = f_j$. We are now going to define a bootstrap procedure for the estimation of the distribution of $\pi(h)$. For this purpose for every $1 \le j \le k$ draw resamples $(X^*_{j,1}, \ldots, X^*_{j,n_j})$ from the set $\{X_{j,1}, \ldots, X_{j,n_j}\}$. Note that this resampling is different from the one proposed in Section 1 because we do not assume $f_1 = \ldots = f_k$. For simplicity we do not indicate this difference in the notation. We define kernel estimates

$$\hat{f}^*_{j,h}(x) = n_j^{-1} \sum_{i=1}^{n_j} K_h\,(X^*_{j,i} - x) \, .$$

Furthermore, we put

$$\pi^*_{ij}(h) = \Big[\int \Big(\hat{f}^*_{i,h}(x) - \hat{f}^*_{j,h}(x) - \hat{f}_{i,h}(x) + \hat{f}_{j,h}(x) \Big)^2 dx^{\;-1/2} - \big(n_i^{-1} + n_j^{-1}\big) K_h^{(2)}(0) \Big]$$

$$\left[2K^{(4)}(0)\, h^{-1} \int \Big(n_i^{-1} \hat{f}^*_{i,h}(x) + n_j^{-1} \hat{f}^*_{j,h}(x) \Big)^2 dx \right]^{-1/2} .$$

The next theorem shows that bootstrap works here. For brevity we consider only the case of a nonrandom bandwidth h_n.

THEOREM 6. *Assume* $h = h_n \to 0$, $n h_n^2 \to \infty$ *and*

(A2') $f_1, \ldots, f_k$ *have compact support and are bounded.*

Furthermore suppose (A1) *and* (A3). *Then for* $1 \le i < j \le k$

$$d_L(\mathcal{L}(\pi_{ij}(h_n)), \; \mathcal{L}^*(\pi^*_{ij}(h_n)) \to 0 \qquad \textit{(in probability)},$$

$$d_L(\mathcal{L}(\pi(h_n)), \; \mathcal{L}^*(\pi^*(h_n)) \to 0 \qquad \textit{(in probability)}.$$

One might be tempted to bootstrap the following statistic instead of $\pi_{ij}(h)$

$$\widetilde{\pi}_{ij}(h) = \int \left(\hat{f}_{i,h}(x) - \hat{f}_{j,h}(x)\right)^2 \, dx - \int \left(\mathcal{K}_h(f_i - f_j)(x)\right)^2 \, dx .$$

But the arguments given in Chapter 2 Section 2 show that bootstrap does not work here (see also the lemma stated there). Note that for $f_i = f_j$ the statistic $\widetilde{\pi}_{ij}(h)$ is asymptotically equivalent to a pure U statistic (see (1.1)) and that therefore - as in Chapter 2 Section 2 - bootstrap is inconsistent on the hypothesis $f_i = f_j$.

3. Proofs

PROOF OF THEOREM 1. The proof of Theorem 1 is straightforward and follows as in Hall (1984). Note that

$$\int \left(\hat{f}_h(x)\right)^2 \, dx = \int \left(\mathcal{K}_h \, f(x)\right)^2 \, dx + o_p(1) \tag{3.1}$$

$$= \int f^2(x) \, dx + o_p(1).$$

This can easily be shown by the calculation of the second moments of this U-statistic.

(3.1) implies that $\Delta_{ij}(h)$ can be approximated by a quadratic form $\widetilde{\Delta}_{ij}(h)$, i.e. $\Delta_{ij}(h) = \widetilde{\Delta}_{ij}(h) + o_p(1)$, where

$$\widetilde{\Delta}_{ij}(h) = [T_{ij}(h) - \left(n_i^{-1} + n_j^{-1}\right) K_h^{(2)}(0)]$$

$$\left(n_i^{-1} + n_j^{-1}\right)^{-1} \left[2 \, K^{(4)}(0) \int f^2(x) \, dx\right]^{-1/2} h^{1/2}.$$

The asymptotic normality of this quadratic form can also be proved with the results of de Jong (1987) (see also the proof of Theorem 1 and (3.3), ..., (3.5) in Chapter 2).

PROOF OF THEOREM 2. For simplicity suppose $k = 2$. Then $n = n_1 + n_2$. We will show

$$n \, \hat{h}_n^{1/2} \left[T_{12}(\hat{h}_n) - \left(n_1^{-1} + n_2^{-1}\right) K_{\hat{h}_n}^{(2)}(0) \right] \tag{3.2}$$

$$- n\, h_n^{1/2} \left[T_{12}(h_n) - \left(n_1^{-1} + n_2^{-1}\right) K_{h_n}^{(2)}(0) \right] \to 0 \quad \text{(in probability)}.$$

An analogous result for the denominator of $\Delta_{12}(h_n)$ and $\Delta_{12}(\widehat{h}_n)$ does then imply the statement of the theorem. We write

$$T_{12}(h) = \sum_{i,j=1}^{n} \kappa_{ij}\, K_h^{(2)}(Z_i - Z_j)$$

where

$$Z_i = X_{1,i} \qquad \text{for } 1 \le i \le n_1,$$

$$Z_{n_1+i} = X_{2,1} \qquad \text{for } 1 \le i \le n_2,$$

$$\kappa_{ij} = \begin{cases} n_1^{-2} & \text{for } 1 \le i,j \le n_1 \\ n_2^{-2} & \text{for } n_1 < i,j \le n \\ -n_1^{-1} n_2^{-1} & \text{else} . \end{cases}$$

Consider the process

$$\eta \to n\, (\eta h_n)^{1/2} \; [T_{12}(\eta h_n) - \left(n_1^{-1} + n_2^{-1}\right) K_{\eta h_n}^{(2)}(0) \big] .$$

For (3.1) it suffices to show for every $C_2 > C_1 > 0$ tightness of this process - restricted to $\eta \in [C_1, C_2]$. Because of the asymptotic normality of this process at a fixed point it suffices to show for $\alpha, \beta \in [C_1, C_2]$ (see Billingsley (1968))

$$n^2 h_n\, E\, [\alpha^{1/2}\, T_{12}(\alpha\, h_n) - \alpha^{1/2}\left(n_1^{-1} + n_2^{-1}\right) K_{\alpha h_n}^{(2)}(0)$$

$$- \beta^{1/2}\, T_{12}(\beta h_n) + \beta^{1/2} \left(n_1^{-1} + n_2^{-1}\right) K_{\beta h_n}^{(2)}(0)\,]^2$$

$$\le \text{const.}\ (\beta - \alpha)^2 .$$

But this is easy. First put

$$V_{ij}(h) = K_h^{(2)}(Z_i - Z_j) ,$$

$$V_i(h) = E\,(V_{ij}(h) \mid Z_i) \quad (\text{for } j \neq i),$$

$$V(h) = E\; V_i(h)\;,$$

$$\widetilde{V}_{ij}(h) = V_{ij}(h) - V_i(h) - V_j(h) + V(h)\;,$$

$$\widetilde{V}_i(h) = V_i(h) - V(h)\;.$$

Then, because $\sum_{i,j=1}^{n} \kappa_{ij} = 0$ for every j, one gets

$$\begin{aligned} T_{12}(h) &= \sum_{i,j=1}^{n} \kappa_{ij}\, V_{ij}(h) \\ &= \sum_{i,j=1}^{n} \kappa_{ij}\, \widetilde{V}_{ij}(h) \\ &= \sum_{i \neq j} \kappa_{ij}\, \widetilde{V}_{ij}(h) + \sum_{i=1}^{n} \kappa_{ii}\, K_h^{(2)}(0) \\ &\quad - 2 \sum_{i=1}^{n} \kappa_{ii}\, \widetilde{V}_i(h) - \left(n_1^{-1} + n_2^{-1}\right) V(h). \end{aligned}$$

Now a simple calculation shows

$$\begin{aligned} &n^2 h_n\, E\, [\alpha^{1/2}\, T_{12}(\alpha\, h_n) - \alpha^{1/2} \left(n_1^{-1} + n_2^{-1}\right) K_{\alpha h_n}^{(2)}(0) \\ &\qquad - \beta^{1/2}\, T_{12}(\beta h_n) + \beta^{1/2} \left(n_1^{-1} + n_2^{-1}\right) K_{\beta h_n}^{(2)}(0)\,]^2 \\ &\leq\; O(h_n)\, \{E\, [\widetilde{V}_{12}(\alpha\, h_n) - \widetilde{V}_{12}(\beta\, h_n)\,]^2 \\ &\qquad + n^{-1}\, E[\widetilde{V}_1(\alpha\, h_n) - \widetilde{V}_1(\beta\, h_n)\,]^2 \\ &\qquad + [V(\alpha\, h_n) - V(\beta\, h_n)\,]^2\, \} \\ &=\; O(h_n)\, E[V_{12}(\alpha\, h_n) - V_{12}(\beta\, h_n)]^2 \\ &=\; O(h_n) \int \left[K_{\alpha h_n}^{(2)}(x - y) - K_{\beta h_n}^{(2)}(x - y)\right]^2\, f(x)\, f(y)\, dx\, dy \end{aligned}$$

$$= O(1) \cdot \int \left[K_\alpha^{(2)}(x) - K_\beta^{(2)}(x)\right]^2 dx$$

$$= O((\beta - \alpha)^2).$$

PROOF OF THEOREM 3. For simplicity suppose again $k = 2$ and $n = n_1 + n_2$. We will show

$$d_L\,(\mathcal{L}(n\, h_n^{1/2}\, T_{12}(h_n)),\ \mathcal{L}^*(n\, h_n^{1/2}\, T_{12}^*(h_n))\,) \to 0 \tag{3.3}$$

(in probability) .

Because of (3.2) and

$$\int \left(\hat{f}_h^*(x)\right)^2 dx \;-\int f^2(x)\; dx = o_p(1), \tag{3.4}$$

this implies the statement of the theorem. (3.4) can easily be shown by the calculation of the second moments of this U-statistic. For the proof of (3.3) we use the strong approximation based on the quantile transformation which is discussed in Section 1 after Theorem 4. We construct $Y_1, \dots, Y_n,\ Y_1^*, \dots, Y_n^*$ as discussed there. Define

$$\tilde{f}_{1,h}(x) = n_1^{-1} \textstyle\sum_{i=1}^{n_1} K_h(Y_i - x)\,,$$

$$\tilde{f}_{2,h}(x) = n_2^{-1} \sum_{i=n_1+1}^{n} K_h(Y_i - x)\,,$$

$$\tilde{f}_{1,h}^*(x) = n_1^{-1} \textstyle\sum_{i=1}^{n_1} K_h\left(Y_i^* - x\right),$$

$$\tilde{f}_{2,h}^*(x) = n_2^{-1} \sum_{i=n_1+1}^{n} K_h\left(Y_i^* - x\right).$$

We consider

$$D(h) = \int \left(\tilde{f}_{1,h}(x) - \tilde{f}_{2,h}(x)\right)^2 dx$$

$$- \int \left(\tilde{f}^*_{1,h}(x) - \tilde{f}^*_{2,h}(x) \right)^2 dx \ .$$

For the proof of (3.3) it suffices to show

$$E^*\left(\left[n\, h_n^{1/2}\, D(h_n) \right]^2 \right) = o_p(1) \ . \tag{3.5}$$

With the same κ_{ij} as in the proof of Theorem 2 one can write

$$D(h) = \sum_{i,j=1}^{n} \kappa_{ij} \left[K_h^{(2)}(Y_i - Y_j) - K_h^{(2)}\left(Y_i^* - Y_j^*\right) \right]$$

Now, as in the proof of Theorem 2, we put

$$W_{ij}(h) = K_h^{(2)}(Y_i - Y_j) - K_h^{(2)}\left(Y_i^* - Y_j^*\right),$$

$$W_i(h) = E^*(W_{ij}(h) \mid Y_i) \ \ (\text{for } j \neq i),$$

$$W(h) = E^* \ W_i(h),$$

$$\widetilde{W}_{ij}(h) = W_{ij}(h) - W_i(h) - W_j(h) + W(h),$$

$$\widetilde{W}_i(h) = W_i(h) - W(h).$$

Then we get

$$D(h) = \sum_{i,j=1}^{n} \kappa_{ij}\, W_{ij}(h)$$

$$= \sum_{i \neq j} \kappa_{ij}\, \widetilde{W}_{ij}(h) - 2 \sum_{i=j}^{n} \kappa_{ij}\, \widetilde{W}_i(h)$$

$$- \left(n_1^{-1} + n_2^{-1} \right) W(h) \ .$$

Therefore

$$E^* \left[n\, h_n^{1/2}\, D(h_n) \right]^2$$

$$= O(h_n) \ E^* \, \widetilde{W}_{12}^{\,2}(h_n)$$

$$= O(h_n)\ E^* [K_{h_n}^{(2)}(Y_1 - Y_2) - K_{h_n}^{(2)}(Y_1^* - Y_2^*)]^2$$

$$= O(h_n) \cdot E^* \{ [(Y_1 - Y_2) - (Y_1^* - Y_2^*)]^2$$

$$[1 (| Y_1 - Y_2 | \le 2h_n) + 1(| Y_1^* - Y_2^* | \le 2h_n)] \}$$

$$\cdot \ \sup_x \left| \frac{\partial}{\partial x} K_{h_n}^{(2)}(x) \right|^2$$

$$= O_p(h_n^{-2})\, E^* (Y_1 - Y_1^*)^2$$

$$= O_p(h_n^{-2}) \int_0^1 \left(F^{-1}(t) - \widehat{F}_n^{-1}(t)\right)^2 dt$$

$$= O_p\left(h_n^{-2}\, n^{-1}\right)$$

where we have used that f is bounded away from 0. This implies (3.5) because of $n h_n^2 \to \infty$.

PROOF OF THEOREM 4. Theorem 4 is an immediate consequence of Theorem 2 and Theorem 3.

PROOF OF THEOREM 5. Assume k = 2. As in the proof of Theorem 3 it suffices to show

$$d_L\Big(\mathcal{L}(n\, \widehat{h}_n^{1/2}\, [T_{12}(\widehat{h}_n) - t_{12}(\widehat{h}_n)]),\ \mathcal{L}^*(n(\widehat{h}_n^{\,*})^{1/2}\, [T_{12}(\widehat{h}_n^{\,*}) - t_{12}(\widehat{h}_n^{\,*})])\Big) = o_p(1)\ , \tag{3.6}$$

where $t_{12}(h) = (n_1^{-1} + n_2^{-1}) K_h^{(2)}(0)$.

For the proof of (3.6) we use the same strong approximation as in the proof of Theorem 3. We define $\tau(h) = \int \left(\tilde{f}_{1,h}(x) - \tilde{f}_{2,h}(x)\right)^2 dx - t_{12}(h)$. For (3.6) it suffices to show

$$n\,h_n^{1/2}\ D(\tilde{h}_n^{\,*}) = o_p(1)\ , \tag{3.7}$$

$$n[(\tilde{h}_n^{\,*})^{1/2} - \tilde{h}_n^{1/2}]\,\tau(\tilde{h}_n) = o_p(1)\ , \tag{3.8}$$

$$n\,h_n^{1/2}\,[\tau(\tilde{h}_n^{\,*}) - \tau(\tilde{h}_n)] = o_p(1)\ , \tag{3.9}$$

where $D(h)$, $\tilde{f}_{1,h}$, $\tilde{f}_{2,h}$ are defined as in the proof of Theorem 3.

We consider the following processes

$$\eta \to n\ h_n^{1/2}\,D(\eta\ h_n)\ ,$$

$$\eta \to n\ h_n^{1/2}\,\tau(\eta\ h_n)\ .$$

In the proof of Theorem 2 we have shown that for every $C_2 > C_1 > 0$ the second process - restricted to $[C_1,C_2]$ - is tight. This implies (3.8) and (3.9). Furthermore with the same arguments as in the proof of Theorem 3 one can show for every $C > 0$

$$n\,h_n^{1/2}\ D(C\ h_n) = o_p(1)\ . \tag{3.10}$$

With (3.10) and with the same arguments as in the proof of tightness in the proof of Theorem 2 one can show tightness of the process $n\,h_n^{1/2}\ D(\eta h_n)$ for $\eta \in [C_1,C_2]$. Then (3.10) implies for every $C_2 > C_1 > 0$

$$\sup_{C_1 \le \eta \le C_2}\ n h_n^{1/2}\ D(\eta h_n) = o_p(1)\ .$$

This shows (3.7).

PROOF OF THEOREM 6. Theorem 6 can be proved along the lines of the proof of Theorem 3.

4. A bootstrap test on the number of modes of a density.

1. Introduction. In this chapter we consider another application of bootstrap in nonparametric density estimation. We present results about the (random) number of modes of a kernel density estimate. These results are applied in the asymptotic treatment of bootstrap tests on the number of modes of a density. The material of this chapter is also contained in Mammen, Marron, and Fisher (1992).

In an exploratory analysis of a data set $X_1, \dots, X_n$ it may be important to decide if a mode of a kernel density estimate is caused by a mode of the underlying density or by random fluctuations of the data. A statistical approach to this problem can be based on hypothesis testing on the number of modes. For a discussion of other multimodality tests see Good and Gaskins (1980), Hartigan and Hartigan (1985), Müller and Sawitzki (1990, 1991) and Silverman (1981, 1983, 1986). For testing the hypothesis of k modes Silverman (1981) has proposed a test based on the "critically smoothed" bandwidth $h = h_{CRIT,k}$ of the kernel density estimator

$$\hat{f}_h(x) = n^{-1} \sum_{i=1}^{n} K((X_i - x) / h) / h ,$$

i.e. the bandwidth h for which the estimator is just between having k and k + 1 modes. A null hypothesis of k modes should be rejected for large values of $h_{CRIT,k}$. The quantity $h_{CRIT,k}$ is uniquely defined if one uses a Gaussian kernel $K = \varphi$ because then the number of modes of $\hat{f}_h$ is monotone in h (see Silverman 1981). In Cuevas and Gonzalez Manteiga (1991) the number of inflection points is considerd and statistical applications of the critical bandwidth according to this number are discussed. It is an appealing feature of these procedures that no bandwidth has to be chosen in advance.

$h_{CRIT,k}$ is the amount of smoothing which is necessary to get a k - modal density. Therefore $h_{CRIT,k}$ measures how far the empirical measure lies away from the class of measures with k - modal densities. Tests of multimodality can be based also on other distances. We would like to mention here the following test statistics T based on L_p norms:

$$T = \int \left(\hat{f}_g - \tilde{f}_k \right)^p .$$

Here g is a bandwidth which has to be chosen in advance. $\tilde{f}_k$ is a density estimate with k modes. Possible choices of $\tilde{f}_k$ are a kernel estimate $\hat{f}_h$ with critical bandwidth $h_{CRIT,k}$ or the density $\tilde{f}_k$ which minimises for a $q \leq 1$

$$\int (\hat{f}_g - \tilde{f}_k)^q .$$

In particular, for p = q we get a minimum distance test. Other curve estimates under monotonicity constraints have been discussed in Mammen (1991a).

In Section 2 we present results about the (random) number of modes of a kernel density estimator. These results can be applied in the asymptotic treatment of the multimodality test of Silverman. For the determination of critical values Silverman(1981) proposed use of a smoothed bootstrap. The asymptotic behaviour of this bootstrap is discussed in Section 3. The proofs are contained in Section 4.

2. The number of modes of a kernel density estimator. In this section we will consider a density f which has exactly j modes. We will assume that f does not lie on the "boundary" of the null hypothesis : f' has no zero crossing x_S of higher order (i.e. $f'(x_S) = f''(x_S) = 0$). For instance this excludes horizontal stretches ($f'(x) = 0$ for x in an interval) and critical points which are neither minima nor maxima. Densities on the "boundary" of the null hypothesis can be treated using similar methods to those in this paper. More formally, we make the following assumptions.

ASSUMPTIONS

(A1) f *is a bounded density with bounded support* [a,b].

(A2) f *is twice continuously differentiable on* (a,b).

(A3) $f'(a+) > 0$, $f'(b-) < 0$.

(A4) f *has* j *local maxima* $z_0 < z_2 < ... < z_{2j-2}$ *and* (j-1) *local minima* $z_1 < ... < z_{2j-3}$ *on* (a,b).

(A5) $f''(x) \neq 0$ *and* $f(x) > 0$ *for all* x *with* $f'(x) = 0$.

In our first theorems we consider the number N(h) of modes of the kernel estimator $\hat{f}_h$ with (nonrandom) bandwidth. In Theorem 1 an asymptotic expansion for the expected number E N(h) of modes will be given for bandwidths $h = h_n$ which are of order $n^{-1/5}$.

THEOREM 1. *For a sequence of bandwidth* $h = h_n$ *assume* $0 < \liminf_{n\to\infty} n^{1/5} h < \limsup_{n\to\infty} n^{1/5} h < \infty$. *Then*

$$E\,N(h) = j + \sum_{p=0}^{2j-2} H\left(\frac{\sqrt{n\,h^5}}{\|\varphi''\|} \frac{|f''(z_p)|}{\sqrt{f(z_p)}} \right) + o(1)$$

where
$$\|\varphi''\| = \left(\int (\varphi'')^2 \right)^{1/2} = (3/8)^{1/2}\ (\pi)^{-1/4},$$

$$H(x) = \varphi(x) / x + \Phi(x) - 1 .$$

The gist of the theorem may be clearer in the following form.

COROLLARY 1.1. *For a constant* c *assume* $h = c\,n^{-1/5}$. *Then*

$$E\,N(h) = j + u(c) + o(1)$$

where u *is a decreasing, continous function with* $\lim_{x\to 0} u(x) = \infty$ *and* $\lim_{x\to\infty} u(x) = 0$.

In Mammen (1991d) it has been shown that the theorem remains valid for a large class of kernels K if one replaces $\|\varphi''\|$ by $\|K''\|$ in the formula. Furthermore there an asymptotic expansion is given for the number of inflection points. One may interpret the additional number of modes or inflection points of $\hat{f}_h$ (compared with f) as a measure of smoothness of $\hat{f}_h$. (This is indeed one of the ways our eyes measure smoothness of a curve.) Then Theorem 1 can be applied to decide how one should choose the kernel K if one wants to get a kernel estimator $\hat{f}_h$ with the same smoothness (i. e. nearly the same number of modes) as f. For related ideas see also Mammen(1991b, 1991c).

Because of the monotonicity of N(h) and E N(h), Theorem 1 has an immediate consequence for the bandwidths larger than $n^{-1/5}$.

COROLLARY 1.2. *Assume* $n^{1/5} h \to \infty$. *Then*

$$E\,N(h) \le j + o(1).$$

The next theorem considers the case of bandwidth $h = h_n$ smaller than $n^{-1/5}$.

THEOREM 2. *If* $h = h_n = o(n^{-1/5})$ *, then* N(h) *tends in probability to* ∞ .

Silverman(1983) showed that on the alternative (i. e. $k < j$) the critical bandwidth $h_{CRIT,k}$ is of order n^0, i. e., there exists a constant $h_0(f,k)$ such that $P(h_{CRIT,k} > h_0(f,k)) \to 1$. Corollary 1.2 and Theorem 2 imply that on the null hypothesis ($k \ge j$), the critical bandwidth is of order $n^{-1/5}$.

COROLLARY 2.1. *Assume* $k \ge j$ *. Then for* $\alpha_n \to 0$, $\beta_n \to \infty$

$$P(\alpha_n n^{-1/5} \le h_{CRIT,k} \le \beta_n n^{-1/5}) \to 1.$$

The rate of convergence of $h_{CRIT,k}$ coincides with the bandwidth which is optimal in most senses for the class of twice continuously differentiable densities. Therefore on the null hypothesis the multimodality test of Silverman is based on the inspection of a reasonable kernel density estimate. In the proofs we will see that $h_{CRIT,k}$ is of order $n^{-1/5}$ because the second derivative $\hat{f}_h''$ is of order $O_P(1)$ if h is of order $n^{-1/5}$. Note that the optimal bandwidth for twice continuously differentiable densities is of the same order $O(n^{-1/5})$ for another reason, namely because of the balance of bias and variance for this order of convergence.

The machinery developed for the proofs of Theorems 1 and 2 easily allows two other interesting results. The next theorem shows that the behavior of the Silverman test is determined by the observations near the local extrema z_i of f.

THEOREM 3. *Assume* $h = h_n$ *is of order* $n^{-1/5}$ (*i. e.* $0 < \liminf_{n\to\infty} n^{1/5} h < \limsup_{n\to\infty} n^{1/5} h < \infty$). *Then for the modes* u_p , $p = 1, 2, \ldots$ *of* $\hat{f}_h$ *the following holds :*

$$\max_p \min_i |u_p - z_i| = O_P(n^{-1/5}) .$$

The next theorem shows that for $k = j$ the asymptotic distribution of $h_{CRIT,k}$ depends only on the values of the density and its second derivative at the points of local extremes.

For this an assumption on the smoothness of f" is added which is slightly stronger than uniform continuity, but weaker than any Lipschitz condition.

THEOREM 4. *Assume that*

$$(\log n)^{3/2} \sup (|f''(t) - f''(z_p)| : |t - z_p| \leq c (\log n)^{1/2} n^{-1/5}) \to 0$$

for every $c > 0$ *and for each* $p = 0, \dots, 2j - 2$. *Then* $n^{1/5}\, h_{CRIT,j}$ *converges weakly to* $V = \sup\limits_{0 \leq p \leq 2j-2} \dfrac{f(z_p)^{1/5}}{|f''(z_p)|^{2/5}} V_p$ *where* $V_0, \dots, V_{2j-2}$ *are i.i.d. variables distributed according to a universal distribution.*

Theorem 1 can be used to obtain critical values of $h_{CRIT,j}$ considered as a test statistic of the null hypothesis of j modes. Given α, we seek for c_α such that

$$\alpha = P(h_{CRIT,j} > c_\alpha) = P(N(c_\alpha) > j).$$

To use Theorem 1 for this, note that

$$E(N(c_\alpha) - j) = T_1 + T_2,$$

where

$$T_1 = E(N(c_\alpha) - j)\, I(N(c_\alpha) < j),$$

$$T_2 = E(N(c_\alpha) - j)\, I(N(c_\alpha) > j).$$

But

$$T_2 = P(N(c_\alpha) > j) + T_3$$

where $T_3 = \sum_{m=1}^{\infty} P(N(c_\alpha) - j > m)$. Consider now the case that the level α is small ($\alpha \to 0$). Then $n^{1/5} c_\alpha \to \infty$ (because of Corollary 2.1) and we conjecture

$$T_3 = O(\, P(N(c_\alpha) > j)^2\,) = O(\alpha^2),$$

and furthermore, if α tends to 0 slowly enough,

$$|T_1| \leq j \, P(\hat{f}_h \text{ has (strictly) less than j modes for } h = c_\alpha) = o(\alpha).$$

Therefore one gets the following approximation

$$\alpha = P(h_{CRIT,j} > c_\alpha) = P(N(c_\alpha) > j) \sim E(N(c_\alpha) - j).$$

So a sensible choice of c_α is the solution of

$$\alpha = \sum_{p=0}^{2j-2} H\left(\frac{\sqrt{n\, c_\alpha^5}}{\|\varphi''\|} \frac{|\widehat{f''(z_p)}|}{\sqrt{\widehat{f(z_p)}}} \right),$$

where $\widehat{f(z_p)}$ (or $\widehat{f''(z_p)}$ respectively) is an estimate of f (of f'' respectively) at the p-th extremum. This requires the estimation of the extremum points and of a second derivative. As an alternative we consider in the next section bootstrap estimates of critical values.

3. Bootstrapping the test statistic. To judge the stochastic behavior of $h_{CRIT,j}$ under the null hypothesis of j modes Silverman proposed using a smoothed bootstrap approach based on resampling from $\hat{f}_h$ where $h = h_{CRIT,j}$. This is a density on the boundary of the hypothesis of j modes. The "critically smoothed" bandwidth for a kernel density estimate in the bootstrap resample may be called $h^*_{CRIT,j}$. Then the bootstrap test rejects the hypothesis that the underlying density has at most j modes if

$$T = P(h^*_{CRIT,j} \leq h_{CRIT,j} \mid X_1, \ldots, X_n) \geq 1 - \alpha .$$

(Actually, Silverman (1981) proposed resampling from a slight modificafion of $\hat{f}_{h_{CRIT,j}}$ instead of from $\hat{f}_{h_{CRIT,j}}$ itself. This modification has an important effect for small sample sizes but is negligible in first order asymptotics.) As alternatives one should also consider resampling from other densities on the "boundary" of the null hypothesis of j modes. One possibility is the L_2 projection of a kernel density estimate onto the densities with j modes (see the introduction and also Mammen (1991a) for the asymptotically equivalent analysis of a related nonparametric regression estimate). We conjecture that the operational characteristics of the Silverman test depend strongly on the choice of the bootstrap density if the true density f itself lies on the "boundary" of the null hypothesis, as in the case that f is constant on a interval. We discuss here neither this case nor the right choice of the bootstrap density. For a discussion of this point and of other modifications of the

Silverman test we refer to Fisher, Mammen and Marron (1990). For other nonparametric curve estimates under the condition of simple curve characteristics see also Mammen (1991b, 1991c).

The next theorem shows that $h^*_{CRIT,j}$, like $h_{CRIT,j}$, is of order $n^{-1/5}$. This suggests that probability $P(T \geq 1-\alpha)$ of rejecting the null hypothesis is asymptotically bounded away from 0 also for the more interesting densities lying in the interior of the null hypothesis.

THEOREM 5. *For* $\alpha_n \to 0$, $\beta_n \to \infty$

$$P(\alpha_n n^{-1/5} \leq h^*_{CRIT,j} \leq \beta_n n^{-1/5} | X_1 , \dots , X_n) \to 1 \quad (\textit{in probability}) .$$

For statistical validity of the bootstrap test the minimal requirement would be that the test has (at least asymptotically) level α , i.e. that

$$(*) \qquad \limsup_{n \to \infty} P(T \geq 1 - \alpha) \leq \alpha$$

for every underlying density with at least j modes. We are unable to establish (*) rigorously, but heuristically. Theorem 4 suggests that the bootstrap distribution $\mathcal{L}(h^*_{CRIT,j} | X_1, \dots , X_n)$ is stochastically larger than $\mathcal{L}(h_{CRIT,j})$, because $| \hat{f}''_{h_{CRIT,j}}(t) |$ can be expected to be of similar size as $|f''(z_p)|$ in most of the z_p - neighborhoods and to be smaller than $|f''(z_p)|$ in one of them (Note that $\hat{f}'_{h_{CRIT,j}}$ is 0 at two different points in the neighborhood of one z_p; see also Fisher, Mammen and Marron, 1990). This motivates (*).

In the next theorem we show that, if one uses the same (nonrandom) amount of smoothing in the real world and in the bootstrap world, then on average, the bootstrapped kernel estimator has more modes than the kernel estimator based on the real observations. This is our second motivation for (*) .

THEOREM 6. *For bandwidths* $h_1 = h_{1,n}$ *and* $h_2 = h_{2,n}$ *which are of order* $n^{-1/5}$ (*i. e.* $h_i^{-1} = O(n^{1/5})$, $h_i = O(n^{-1/5})$) *consider the kernel estimator* $\hat{f}^*_{h_1}$ *with bandwidth* h_1 *constructed by* n *observations drawn from the density* $\hat{f}_{h_2}$. *Denote the number of modes of the kernel estimator* $\hat{f}^*_{h_1}$ *in an interval* (s_n, t_n) *by* $N^*(h_1, s_n, t_n)$. *Assume that* s_n *and* t_n *do not lie in an* $n^{-1/5}$ *neighborhood of a local extremum of* f

$$n^{1/5} \min_{0 \le p \le 2j-2} |s_n - z_p| \to \infty,$$

$$n^{1/5} \min_{0 \le p \le 2j-2} |t_n - z_p| \to \infty.$$

Then

$$E\, N^*(h_1, s_n, t_n) \geq E\, N(h_1, s_n, t_n) + o(1),$$

where $N(h_1, s_n, t_n)$ *denotes the number of modes of the kernel estimator* $\hat{f}_{h_1}$ *in the interval* (s_n, t_n).

Note that we do not need to assume $h_1 \le h_2$ or the like. The theorem is still valid even when $h_1 = 100\, h_2$. The reason is that the added smoothness coming from large values of h_1 is automatically inherited by resampling from $\hat{f}_{h2}$.

The conjecture (*) is also supported by the results of a small simulation study. The underlying densities were three unimodal densities

a) *the standard normal,*

b) *the convolution of the standard normal and of the uniform distribution on* (0, 0.5),

c) *the convolution of two uniform distributions on* (0, 1).

For each of 100 runs a pseudo random sample of 200 observations was generated. For each data set bootstrap P - values (i.e. 1 - T) were calculated. In figures 1 and 2 a Q - Q plot of the bootstrap P - values against uniform quantiles (i - 0.5)/100 is plotted. Figure 1 shows the Q - Q plot for the test of the null hypothesis of one mode and figure 2 for the null hypothesis of two modes. Note that every plot lies above the diagonal for α smaller than 0.5. This implies that the test is indeed conservative in these three cases as conjectured by (*). For further simulations and for a more detailed discussion we refer to Fisher, Mammen and Marron (1990).

In the here described resampling bootstrap samples are drawn according to a density $g = \hat{f}_h$ where $\hat{f}_h$ is a kernel estimate with bandwidth $h = h_{CRIT,j}$ of order $n^{-1/5}$. g'' does not estimate consistently f''. However, according to Theorem 4 the asymptotic distribution of $h_{CRIT,j}$ depends on the values of f'' (and f) at local extremes. This is the reason why bootstrap does not work as an estimate of the distribution of $h_{CRIT,j}$. Oversmoothing is a simple way out of it. For this choose a sequence $\gamma_n \to \infty$ with $\gamma_n n^{-1/5} \to 0$. Put $h = \gamma_n h_{CRIT,j}$. Then one can show that $\hat{f}''_h \to f''$ holds uniformly in $n^{-1/5}$ - neighborhoods of local extremes. Consider now resampling from $\hat{f}_h$ with $h = \gamma_n h_{CRIT,j}$. For this resampling one can show with the arguments given in the proofs of Theorems 4 and 5 that the distribution of $h_{CRIT,j}$ is consistently estimated. However, the choice of γ_n remains a difficult practical problem.

4. Proofs. We will give explicit proofs only for the case that f has j = 1 mode. Other cases are very similar, but require additional cumbersome notation. We will make repeated use of the following result in Section 13.2 of Cramér - Leadbetter (1967).

THEOREM (CRAMÉR & LEADBETTER). *For a differentiable Gaussian process* Z(t) *consider the number* N *of zero crossings of* Z *in the interval* [u,v] *. Then*

$$E\,N = \int_u^v \gamma(t)\,\sigma(t)^{-1}\,\rho(t)\;\varphi\Big(\frac{m(t)}{\sigma(t)}\Big)\,G(\eta(t))\,dt$$

where

$$\gamma^2(t) = \mathrm{var}\,(Z'(t)),$$

$$\sigma^2(t) = \mathrm{var}\,(Z(t)),$$

$$\mu(t) = \mathrm{cov}\,(Z(t)\,,\,Z'(t))\,/\,(\gamma(t)\,\sigma(t)),$$

$$\rho(t) = (1-\mu^2(t))^{1/2},$$

$$m(t) = E\,Z(t),$$

$$\eta(t) = \frac{m'(t) - \gamma(t)\,\mu(t)\,m(t)\,/\,\sigma(t)}{\gamma(t)\;\rho(t)},$$

$$G(x) = 2\,\varphi(x) + x\,(2\Phi(x) - 1).$$

PROOF OF THEOREM 1. For $X_n(t) = X_n(t,h) = \hat{f}_h(t) - E\,\hat{f}_h(t)$ there exists a Brownian bridge W^0 (which is independent of h) such that for

$$Y_n(t) = Y_n(t,h) = n^{-1/2} \int h^{-2} \varphi'((x-t)/h)\ W^0(F(x))\, dx$$

$$= n^{-1/2} \int h^{-1} \varphi((x-t)/h)\, dW^0(F(x))$$

the following holds if c_1 is large enough :

$$\text{(4.1)} \qquad P(B_s) = o(n^{-1})$$

where

$$\text{(4.2)} \qquad B_s = \{\sup_t |\, X_n^{(s)}(t) - Y_n^{(s)}(t)\, | \geq c_1 (\log n)\, n^{-(3-s)/5}\}$$

for $0 \leq s \leq 3$. This can be seen for instance as in the proof of Proposition 1 in Silverman (1978). For a fixed number c_0 denote the number of modes of $\widetilde{Y}_n = E\,\hat{f}_h + Y_n$ in the interval $[z_0 - c_0 \sqrt{\log n}\, n^{-1/5}\ ,\ z_0 + c_0 \sqrt{\log n}\, n^{-1/5}]$ by $\widetilde{N}(h)$.

The following three lemmas imply the statement of Theorem 1 (for $j = 1$) .

LEMMA 1. $\quad E\,\widetilde{N}(h) = 1 + H\left(\frac{\sqrt{n h^5}}{\|\varphi''\|} \frac{|f''(z_0)|}{\sqrt{f(z_0)}} \right) + o(1).$

LEMMA 2. $\quad P(\widetilde{N}(h) \neq N(h)) = 0\, ((\log n)^2\, n^{-1/5}).$

LEMMA 3. $\quad$ (i) $\quad E\, N(h)\, I(N(h) > n^{1/5} (\log n)^{-2}) = o(1),$

(ii) $\quad E\,\widetilde{N}(h)\, I(\widetilde{N}(h) > n^{1/5} (\log n)^{-2}) = o(1).$

PROOF OF LEMMA 1. We apply the Cramér-Leadbetter formula for

$$Z(t) = E\,\hat{f}_h'(t) + Y_n'(t) \qquad \text{and}$$

$$u = z_0 - \sqrt{\log n}\, n^{-1/5}\ ,\ \ v = z_0 + \sqrt{\log n}\, n^{-1/5}.$$

Note that for two functions g_1, g_2 one gets

(4.3) $$E \int g_1(x)\, d\, W^0(F(x)) \int g_2(x)\, d\, W^0(F(x))$$

$$= \int g_1(x)\, g_2(x)\, f(x)\, dx \;-\; \int g_1(x)\, f(x)\, dx \int g_2(x)\, f(x)\, dx .$$

This implies for $u \le t \le v$

$$\sigma^2(t) = n^{-1} h^{-4} \int (\varphi'((x-t) / h))^2 f(x)\, dx$$

$$- [n^{-1/2} h^{-2} \int \varphi'((x-t) / h)\, f(x)\, dx]^2$$

$$= n^{-1} h^{-3} \int (\varphi'(x))^2 f(t + hx)\, dx$$

$$- [n^{-1/2} h^{-1} \int \varphi'(x)\, f(t + hx)\, dx]^2 .$$

Now $f'(z_0) = 0$ implies

$$|f(t + hx) - f(z_0)| \le \text{const.}\ [(t - z_0)^2 + h^2 x^2].$$

Putting this into the formula for $\sigma^2(t)$ gives

$$\sigma^2(t) = n^{-1}\ h^{-3}\ ||\varphi'||^2 f(z_0) + 0\ (n^{-1} h^{-3} \cdot n^{-2/5} \log n).$$

Similarly one can show for $u \le t \le v$

$$\gamma^2(t) = n^{-1} h^{-5}\ ||\varphi''||^2 f(z_0) + 0\ (n^{-2/5} \log n),$$

$$\mu(t) = 0\ (n^{-2/5} \log n),$$

$$\rho(t) = 1 + 0\ (n^{-4/5} (\log n)^2),$$

$$m(t) = E\, \hat{f}'_h(t) = f''(z_0)\, (t - z_0) + o\ (\sqrt{\log n}\ n^{-1/5}),$$

$$m'(t) = f''(z_0) + o\ (1).$$

Putting these into the Cramér-Leadbetter formula one gets for the expected number E N of zero crossings of the process $Z(t) = E\, \hat{f}'_h(t) + Y'_n(t)$ in the interval $[u,v] = [\, z_0 - \sqrt{\log n}\ n^{-1/5}\ ,\, z_0 + \sqrt{\log n}\ n^{-1/5}\,]$

(4.4) $$E\,N = \int_u^v \gamma(t)\,\sigma(t)^{-1}\,\varphi(m(t)/\sigma(t))\,G\,(\,m'(t)/\gamma(t)\,)\,dt + o\,(1)$$

$$= \gamma(z_0)/f''(z_0)\ G\,(f''(z_0)/\gamma(z_0)\,) + o\,(1)$$

$$= 1 + 2H\left(\frac{\sqrt{n\,h^5}}{\|\varphi''\|}\ \frac{|f''(z_0)|}{\sqrt{f(z_0)}}\right) + o\,(1)\,.$$

Now use $N = 2\,\tilde{N}\,(h) - 1$.

For the proofs of the Lemmas 2 and 3 we use the following lemmas.

LEMMA 4. *For every* $a_n = o\,(1)$ *and* c_0 :

(4.5) $$P\,(\text{ There exists a } t \in U_n\text{, such that } |\tilde{Y}_n''(t)| \le a_n \text{ and }$$

$$|\tilde{Y}_n'(t)| \le n^{-1/5}\,a_n\,) = 0\,((\log n)\,a_n\,) + o\,(n^{-1}).$$

Here $\tilde{Y}_n(t) = E\,\hat{f}_h(t) + Y_n(t)$ *and* $U_n = \{t : |t - z_0| \le c_0\sqrt{\log n}\ n^{-1/5}\ \}$.

The proof of Lemma 4 depends on the following two lemmas:

LEMMA 5. *For* c *large enough and* $0 \le s \le 3$:

(4.6) $$P\,(A_s) = o\,(n^{-1})$$

where

$$A_s = \{\sup_{a \le t \le b} |Y_n^{(s)}(t)| > c\,\sqrt{\log n}\ n^{-(2-s)/5}\}$$

Lemma 5 and (4.1) imply the following result.

LEMMA 6. *For* c *large enough and* $0 \le s \le 3$

(4.7) $$P\,(\,\sup_{a \le t \le b} |X_n^{(s)}\,(t)| > c\,\sqrt{\log n}\ n^{-(2-s)/5}\,) = o\,(n^{-1})\,.$$

PROOF OF LEMMA 4. Define

$$S = \{t \mid t \in n^{-1/5} a_n (\log n)^{-1/2} \mathbb{Z}\} \cap U_n .$$

Note that $|h\, \delta'''|$ is bounded for $\delta = E\, \hat{f}_h$ and that $\tilde{Y}_n'(t)$ and $\tilde{Y}_n''(t)$ are independent for fixed t (see (4.3)). Then (4.6) implies for c large enough.

$$P(\exists t \in U_n : |\tilde{Y}_n''(t)| \le a_n ,\ |\tilde{Y}_n'(t)| \le n^{-1/5} a_n)$$

$$\le P(\exists s \in S : |\tilde{Y}_n''(s)| \le (c+1)\, a_n ,\ |\tilde{Y}_n'(s)| \le (c+1)\, n^{-1/5} a_n) + o(n^{-1})$$

$$\le \#S \sup_{s \in S} P(|\tilde{Y}_n''(s)| \le (c+1)\, a_n)\, P(|\tilde{Y}_n'(s)| \le (c+1)\, n^{-1/5} a_n) + o(n^{-1})$$

$$\le 0\left(\frac{\log n}{a_n} \cdot a_n^2\right) + o(n^{-1}) = 0((\log n) \cdot a_n) + o(n^{-1}).$$

PROOF OF LEMMA 5. $P(A_s)$ is bounded by the expected number of crossings of the levels $\pm c \sqrt{\log n}\, n^{-(2-s)/5}$ by $Y_n^{(s)}(t)$. These expectations can easily be estimated by the Cramér Leadbetter formula.

We come now to the proofs of Lemmas 2 and 3 .

PROOF OF LEMMA 2. Choose c_0 , c large enough. Then using Lemma 6 one can show that P(There exists a 0-crossing of $\hat{f}_h'$ outside $U_n = [z_0 - c_0 \sqrt{\log n}\, n^{-1/5}, z_0 + c_0 \sqrt{\log n}\, n^{-1/5}]) = o(n^{-1})$.

Applying Lemma 4 with $a_n = c(\log n)\, n^{-1/5}$, we see that with probability 1 - $0((\log n)^2\, n^{-1/5})$, the condition

$$|\tilde{Y}_n''(t)| > c(\log n)\, n^{-1/5} \text{ if } |\tilde{Y}_n'(t)| \le c(\log n)\, n^{-2/5}$$

holds for all $t \in U_n$. Now (4.1) and (4.2) imply, that with probability 1 - $0((\log n)^2\, n^{-1/5})$ the statement

$$\text{if } |\tilde{Y}_n'(t)| \le c(\log n)\, n^{-2/5} \text{ or } |\hat{f}_h'(t)| \le (c + c_1)(\log n)\, n^{-2/5}$$

then $|\widetilde{Y}_n''(t)| > c\,(\log n)\,n^{-1/5}$ and $|\hat{f}_h''(t)| > (c - c_1)\,(\log n)\,n^{-1/5}$

holds for all $t \in U_n$. In other words

$\widetilde{Y}_n'$ and $\hat{f}_h'$ are monotone if $|\widetilde{Y}_n'|$ or $|\hat{f}_h'|$ are small.

This implies the statement of Lemma 2 .

PROOF OF LEMMA 3. To prove (i) we first show that

$$N(h) \leq n.$$

This follows from

$$\lim_{h \to 0} N(h) \leq n$$

and from the fact that N(h) is decreasing. The statement (i) of Lemma 3 follows now by replicated use of the following lemma. The proof of Lemma 3 (ii) follows similarly. First one considers the values of $\widetilde{Y}_n' = E\,\hat{f}_h' + Y_n'$ in n equidistant points in the interval $[z_0 - c_0\sqrt{\log n}\;n^{-1/5}\;,\; z_0 + c_0\sqrt{\log n}\;n^{-1/5}]$. Define the process $\overline{Z}(t)$ as the linear interpolation of these values. Define $\widetilde{N}_n(h)$ as the number of zero downcrossings of $\overline{Z}(t)$. Then $\widetilde{N}_n(h) \leq n$ and $\widetilde{N}_n(h) \leq \widetilde{N}(h)$ (see Section 10.3 in Cramér - Leadbetter (1967)). Now along the lines of the calculations of page 286 in Cramér - Leadbetter (1967) one can show that $E\;\widetilde{N}(h) - \widetilde{N}_n(h) = o(\,(\log n)^2\,n^{-1/5}\,)$. Therefore it suffices to show $E\,\widetilde{N}_n(h)\;I(\widetilde{N}_n(h) > n^{1/5}\,(\log n)^{-2}) = o\,(1)$. But this can be done with similar arguments as in the proof of statement (i).

LEMMA 7. *For every* $b_n \to 0$ *with* $b_n\,n^{\delta} \to \infty$ *for a* $\delta > 0$ *the following holds for* c' *large enough*

$$P\,(N(h) \geq c'\;(\log n)^{5/4}\;/\;\sqrt{b_n}\,) = 0\,(b_n\,) + o\,(n^{-1}). \tag{4.8}$$

PROOF OF LEMMA 7. Write $g = \hat{f}_h'$. First note that Lemma 4 remains valid for $a_n = b_n/(\log(n))$ if we replace $(\,\widetilde{Y}_n''\,,\,\widetilde{Y}_n)$ by $(\,\hat{f}_h''\,,\,\hat{f}_h'\,)$. This can be seen along the lines of the proof of Lemma 4. The only point which needs a closer look is

$$P(|\hat{f}_h''(s)| \leq (c+1)\, a_n \,,\, |\hat{f}_h'(s)| \leq (c+1)\, n^{-1/5}\, a_n) = 0\left(a_n^2\right).$$

But this follows from the validity of a higher - order Edgeworth expansion for $\left(n^{1/5}\hat{f}_h'(s),\, \hat{f}_h''(s)\right)$. The validity of a higher - order Edgeworth expansion can be proved with the same arguments as in Hall(1990a). Note that with a truncation argument and the same arguments as in the proof of Lemma 4.1 in Hall(1990a) one can show for the normal kernel $K = \varphi$ the following (Cramér type) condition:

$$\sup_{|t_1|+|t_2|>\varepsilon} \left| \int_{-\infty}^{\infty} \exp(\, i\, t_1 K'(u) + i t_2\, K''(u))\, h\, f(x-hu)\, du \right| \leq 1 - C(\varepsilon)\, h$$

for a constant $C(\varepsilon) > 0$ depending on ε.

Therefore one gets with probability $1 - 0\,(b_n) - o\,(1/n)$:

After every 0 crossing $g = \hat{f}_h'$ is monotone until it reaches the level $n^{-1/5}\; b_n / \log n$ or the level $- n^{-1/5}\, b_n / \log n$.

Furthermore for c large enough, because of Lemma 6 the following holds with probability $1 - o\,(n^{-1})$:

$$|g''| \leq c\, \sqrt{\log n}\; n^{1/5} \;.$$

The minimal distance of two 0 crossings of a function g fulfilling these two conditions is

$$2\, \sqrt{b_n}\; n^{-1/5}\, (\log n)^{-3/4}\, c^{-1/2}.$$

This can be seen as follows. Consider a zero crossing t_0 where g is monotone increasing. Define

$$\delta = \inf\{t > 0 \mid g(t_0 + t) \geq n^{-1/5}\, b_n / \log n \,;\, g'(t_0 + t) = 0\},$$

$$\Delta = \sup\{g'(t) \mid t_0 \leq t \leq t_0 + \delta\} \,.$$

Then:

$$\delta \geq \frac{\Delta}{c\sqrt{\log n}} n^{-1/5}$$

and

$$\delta \cdot \Delta \geq n^{-1/5} b_n / \log n.$$

But this implies $\delta \geq \sqrt{b_n}\, n^{-1/5} (\log n)^{-3/4} c^{-1/2}$.

The statement of Lemma 7 now follows because with probability $1 - o(n^{-1})$ there is no zero - crossing of $\hat{f}_h'$ outside of $U_n = [z_0 - c_0 \sqrt{\log n}\, n^{-1/5}, z_0 + c_0 \sqrt{\log n}\, n^{-1/5}]$.

Therefore

$$0(b_n) + o(1/n)$$

$$= P(N(h) \geq [2 c_0 \sqrt{\log n}\, n^{-1/5}] / [2 \sqrt{b_n}\, n^{-1/5} (\log n)^{-3/4} c^{-1/2}] |$$

$$= P(N(h) \geq c_0 c^{1/2} (\log n)^{5/4} / \sqrt{b_n}) .$$

PROOF OF THEOREM 2. Because N(h) is decreasing in h we can assume $n^{-1} h^{-2} (\log n) = o(n^{-1/5})$. Put $I_{j,n} = [(j-1) n^{-1/5} + z_0 , j n^{-1/5} + z_0]$ for $j \in \mathbb{Z}$. Proceeding as in the proof of Proposition 2 in Silverman (1983) one can show for every $j \in \mathbb{Z}$:

$$n^{1/5} \sup_{x \in I_{j,n}} \hat{f}_h'(x) \to \infty \quad \text{(in prob.)}$$

$$n^{1/5} \inf_{x \in I_{j,n}} \hat{f}_h'(x) \to -\infty \quad \text{(in prob.)}$$

This proves Theorem 2 . Proposition 2 in Silverman (1983) contains a stronger statement than Theorem 2, but the proof is defective. The first equation on page 257, and therefore also (18), does not follow from (17) and (15). (In (17) the exponent 1/2 should be replaced by -1/2 .) Therefore our Corollary 2.1 corrects a result in Silverman(1983) where was erroneously stated that on the null hypothesis the critical bandwidth is of order

$n^{-1/5} \log(n)^{1/5}$.

PROOF OF THEOREM 3. Consider the Gaussian process Y_n constructed in the beginning of the proof of Theorem 1 . Choose $\Delta_n \to \infty$, $\Delta_n = o(\sqrt{\log n})$. For c_2 large denote the number of 0 crossings of $Y_n'(t) + E\hat{f}_h'(t) + c_1(\log n)\, n^{-1/5}$ in the interval $[z_0 + \Delta_n \cdot n^{-1/5}, z_0 + c_2 \sqrt{\log n}\, n^{-1/5}]$ by $N_1(h)$. Similarly $N_2(h)$ is the number of 0 crossings of $Y_n'(t) + E\hat{f}_h'(t) - c_1(\log n)\, n^{-1/5}$ in the interval $[z_0 - c_2 \sqrt{\log n}\, n^{-1/5}, z_0 - \Delta_n n^{-1/5}]$. The theorem follows from

$$P(N_j(h) > 0) \leq E N_j(h) \to 0 .$$

Note that $P(|Y_n'(t) + E\hat{f}_h'(t) - \hat{f}_h'(t)| > c_1 (\log n)\, n^{-1/5}) \to 0$. The convergence E $N_j(h)$ follows by an application of the Cramér-Leadbetter formula. Note for instance (see (4.4)) that with a sequence s_n bounded away from 0 and ∞

$$E N_1(h) = \int_{z_0+\Delta_n n^{-1/5}}^{z_0+c_2\sqrt{\log n}\, n^{-1/5}} \sigma(t)^{-1} \varphi(m(t)/\sigma(t))\, dt\, (1 + 0(1))$$

$$= \int_{z_0+\Delta_n n^{-1/5}}^{z_0+c_2\sqrt{\log n}\, n^{-1/5}} n^{1/5} s_n^{-1} \varphi(m(t)\, n^{1/5} s_n^{-1})\, m'(t)\, dt\, (1 + 0(1))$$

$$= \int_{t_1}^{t_2} \varphi(t)\, dt\, (1 + 0(1))$$

where $t_1 = n^{1/5} s_n^{-1} m\left(z_0 + \Delta_n n^{-1/5}\right)$ and $t_2 = n^{1/5} s_n^{-1} m\left(z_0 + c_2 \sqrt{\log n}\, n^{-1/5}\right)$. Now $E N_1(h) \to 0$ follows from $t_1 \to \infty$.

PROOF OF THEOREM 4. As in the beginning of the proof of Theorem 1 define $Y_n(t,h)$. Choose m fixed constants $0 < \delta_1 < \ldots < \delta_m$. Put $h_i = \delta_i \cdot n^{-1/5}$. Put

$$h_{CRIT,j,1} = \inf\{h_i \mid \hat{f}_h \text{ has exactly one local maximum for } h = h_i\},$$

$$h_{CRIT,j,2} = \inf\{h_i \mid E\hat{f}_h(t) + Y_n(t,h) \text{ has exactly one local maximum for } h = h_i\},$$

$$h_{CRIT,j,3} = \inf\{h_i \mid \tfrac{1}{2} f''(z_0)(t - z_0)^2 + Y_n(t,h) \text{ has exactly one local maximum for}$$

$h = h_i\}$,

where the infimum over an empty set is defined as $+\infty$.

We will show that

$$P(h_{CRIT,j,1} = h_{CRIT,j,3}) \rightarrow 1.$$

This shows that the distribution of $h_{CRIT,j}$ has the same weak limit as the distribution of the critical bandwidth h of $\frac{1}{2} f''(z_0)\,(t - z_0)^2 + Y_n(t,h)$. This limit distribution depends only on $f(z_0)$ and $f''(z_0)$. This proves the theorem.

Now $P(h_{CRIT,j,1} = h_{CRIT,j,2}) \rightarrow 1$ follows from Lemma 2 and Theorem 3. So it remains to show $P(h_{CRIT,j,2} = h_{CRIT,j,3}) \rightarrow 1$. But this follows from Lemma 4 if one chooses

$$a_n = 2\sqrt{\log n} \cdot (c_0 \vee 1)\, \sup\{|f''(t) - f''(z_0)|: \ |t - z_0| \le \sqrt{\log n}\, c_0\, n^{-1/5}\}$$

and uses

$$|E\, \hat{f}'_h(t) - f''(z_0)\,(t - z_0)| \le \frac{a_n}{2} \cdot n^{-1/5} + o(a_n\, n^{-1/5})$$

$$|E\, \hat{f}''_h(t) - f''(z_0)| \le \frac{a_n}{2} + o(a_n)$$

for $t \in U_n$ (see also the proof of Lemma 2).

PROOF OF THEOREM 5. Consider a resample $X_1^*, \ldots, X_n^*$ with density $\hat{f}_{h_{crit,j}}$. As in the proof of Theorem 1 we use a Brownian bridge W_n^* (given $X_1, \ldots, X_n$) such that for

$$Y_n^*(t,h) = n^{-1/2} \int h^{-1}\, \varphi\,((x-t) / h)\, dW_n^*\,(F_n(x))$$

where F_n is the distribution function with density $\hat{f}_{h_{crit,j}}$, the following holds if c_1 is large enough:

(4.9) $$P(P^*(B^*_{s,h}) \geq n^{-2}) \rightarrow 1$$

where

$$P^*(\ ) = P(\ |X_1, \dots, X_n),$$

$$E^*(\ ) = E(\ |X_1, \dots, X_n),$$

$$B^*_{s,h} = \{ \sup_t |\frac{d^s}{(dt)^s}(\hat{f}^*_h(t) - E^*\hat{f}^*_h(t) - Y^*_n(t,h))| \geq c_1(\log n)\, n^{-(3-s)/5} \}$$

for $0 \leq s \leq 3$.

Denote the number of modes of $\hat{f}^*_h$ by $N^*(h)$. We will show:

(4.10) $$P^*(N^*(h_n) > 1) \rightarrow 1 \qquad \text{(in probability) if } h_n = o(n^{-1/5})$$

(4.11) $$P^*(N^*(h_n) = 1) \rightarrow 1 \qquad \text{(in probability) if } h_n^{-1} = o(n^{1/5}).$$

(4.10) can be shown as in the proof of Theorem 2. To prove (4.11) we proceed as in the proof of Theorem 1. With the arguments given in the proof of Lemma 2 one can show for c_0 large enough

(4.12) $$E\,P^*(\tilde{N}^*(h_n) \neq N^*(h_n)) \rightarrow 0$$

where $\tilde{N}^*(h_n)$ are the number of modes of $Y^*_n(t,h_n) + E^*\hat{f}^*_{h_n}(t)$ in the interval $(u_n, v_n) = (z_0 - c_0\sqrt{\log n}\, n^{-1/5}, z_0 + c_0\sqrt{\log n}\, n^{-1/5})$. So it remains to show that

(4.13) $$E^*\tilde{N}^*(h_n) \rightarrow 1 \qquad \text{(in prob.)}.$$

Proof of (4.13). First note that

$$E^*\hat{f}^*_{h_n} = \hat{f}_\tau \qquad \text{where } \tau = \tau_n = \sqrt{h_n^2 + h_{crit,j}^2}$$

Put $$m(t) = m_n(t) = \hat{f}'_\tau(t),$$

$$\sigma^2 = \sigma_n^2 = n^{-1} h_n^{-3} \|\varphi'\|^2 f(z_0),$$

$$\gamma^2 = \gamma_n^2 = n^{-1} h_n^{-5} \|\varphi''\|^2 f(z_0) .$$

Then by application of the Cramér-Leadbetter formula one gets with similar expansions as in the proof of Theorem 1:

$$\text{(4.14)} \qquad E^* \, 2\, \tilde{N}^*(h_n) - 1$$

$$= \int_{u_n}^{v_n} \gamma \sigma^{-1} \; \varphi(m(t) / \sigma) \, G\, (m'(t) / \gamma) \, dt + o_p(1)$$

$$= U_1 + \ldots + U_4 + o_p\,(1)$$

where

$$U_1 = \int_{u_n}^{v_n} \sigma^{-1} \varphi\left(\frac{m(t)}{\sigma}\right) \cdot 2\gamma\, \varphi\left(\frac{m'(t)}{\gamma}\right) dt,$$

$$U_2 = \int_{u_n}^{v_n} \sigma^{-1} \varphi\left(\frac{m(t)}{\sigma}\right) \; 2\, m'(t)\, \Phi\left(\frac{m'(t)}{\gamma}\right) dt,$$

$$U_3 = \int_{u_n}^{v_n} \sigma^{-1} \varphi\left(\frac{m(t)}{\sigma}\right) (E\, m'(t) - m'(t))\, dt\,,$$

$$U_4 = \int_{u_n}^{v_n} \sigma^{-1} \varphi\left(\frac{m(t)}{\sigma}\right) \; (-E\, m'(t)\,)\, dt\,.$$

Now the theorem is proved by the following statements:

$$\text{(4.15)} \qquad E|U_p| \to 0 \quad \text{for } p = 1, 2, 3,$$

$$\text{(4.16)} \qquad E\, U_4 \;\to 1.$$

For the proof of (4.15) and (4.16) note that m(t) and m'(t) are asymptotically independent. This implies for $p = 1$:

$$E\,|U_1| = \int_{u_n}^{v_n} E\, \sigma^{-1}\, \varphi\left(\frac{m(t)}{\sigma}\right) E\, 2\gamma\, \varphi\left(\frac{m'(t)}{\gamma}\right) dt\,.$$

Now $\gamma \to 0$ (because we have assumed $h_n^{-1} = o(n^{1/5})$) and therefore

$$E\ \sigma^{-1}\, \varphi\left(\frac{m(t)}{\sigma}\right) = \tilde{\sigma}^{-1}\ \varphi\left(E\, m(t) / \tilde{\sigma}\right) + o(1)$$

where $$\tilde{\sigma}^2 = \tilde{\sigma}_n^2 = \sigma^2 + n^{-1}\, \tau_n^{-3}\, \|\varphi'\|^2\, f(z_0)$$

For the proof of the other statements in (4.15), (4.16) use $E(m'(t) - E\, m'(t))^2 \to 0$.

PROOF OF THEOREM 6. Suppose for simplicity that the interval [s,t] coincides with the support [a,b] of f. Write $N^*(h_1) = N^*(h_1, s_n, t_n)$ and, as above, $N(h_1) = N(h_1, s_n, t_n)$. Consider a resample $X_1^*, \ldots, X_n^*$ with density $\hat{f}_{h_2}$. As in the last proof we will use a Brownian bridge W_n^* (given $X_1, \ldots, X_n$) such that for $Y_n^*(t,h) = n^{-1/2} \int h^{-1}\, \varphi((x-t)/h)\, dW_n^*(F_n(h_2, x))$, property (4.9) holds for c_1 large enough. Here $F_n(h_2, x)$ is the distribution function with density $\hat{f}_{h_2}$. We will show:

$$E\, \tilde{N}^*(h_1) \geq E\, N(h_1) + o(1)$$

where, as above, $\tilde{N}^*(h_1)$ is the number of modes of $Y_n^*(t,h_1) + E^*\, \hat{f}^*_{h_1}(t)$ in the interval $(u_n, v_n) = (z_0 - c_0 \sqrt{\log n}\; n^{-1/5},\ z_0 + c_0 \sqrt{\log n}\; n^{-1/5})$. This shows the statement of the theorem because $\lim_{c \to \infty} E \min(\tilde{N}^*(h_1), c) = E\, \tilde{N}^*(h_1)$.

Now put $\tau = \sqrt{h_1^2 + h_2^2}$ and define $m(.)$, σ, γ as in the last proof. Then one can show

$$E\, \tilde{N}^*(h_1) = E\, E^*\, \tilde{N}^*(h_1)$$

$$= 1 + E \int_{u_n}^{v_n} \sigma^{-1}\, \varphi(m(t)/\sigma)\, \frac{\gamma}{2}\, G\left(\frac{m'(t)}{\gamma}\right) dt + o(1)$$

$$= 1 + \int_{u_n}^{v_n} \tilde{\sigma}^{-1}\, \varphi\left(E\, \frac{m(t)}{\tilde{\sigma}}\right) \tilde{\Gamma}\left(E\, m'(t)\right) dt + o(1)$$

where $\tilde{\sigma}$ is defined as in the last proof and where

$$\tilde{\Gamma}(u) = \frac{\gamma}{2} \int G\left(\frac{v}{\gamma}\right) \frac{1}{\bar{\sigma}} \varphi\left(\frac{v-u}{\bar{\sigma}}\right) dv$$

and
$$\bar{\sigma}^2 = n^{-1} \tau^{-5} \|\varphi''\|^2 f(z_0)$$

$$(= \text{var } m'(t) + 0(n^{-2/5} \log n)).$$

Now by differentiation one can see that G is convex. This implies that

$$\tilde{\Gamma}(u) \geq \frac{\gamma}{2} G\left(\frac{u}{\gamma}\right).$$

Using this and $E\, m'(t) = f''(z_0) + o(1)$ gives

$$E\, \tilde{N}^*(h_1) \geq 1 + \int_{u_n}^{v_n} \tilde{\sigma}^{-1} \varphi\left(\frac{E\, m(t)}{\tilde{\sigma}}\right) \frac{\gamma}{2} G\left(\frac{E\, m'(t)}{\gamma}\right) dt + o(1)$$

$$= 1 + \frac{\gamma}{2 f''(z_0)} G\left(\frac{f''(z_0)}{\gamma}\right) + o(1)$$

$$= E\, N(h_1) + o(1).$$

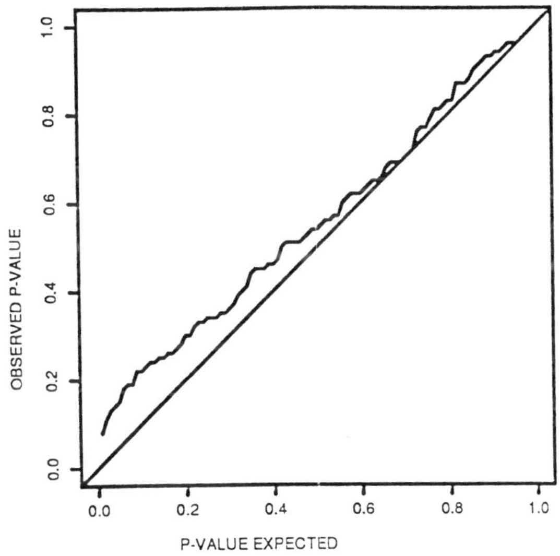

Figure 1a. *Q - Q plot of bootstrap P - values (Test for one mode, observations distributed according to a convolution of two uniforms)*

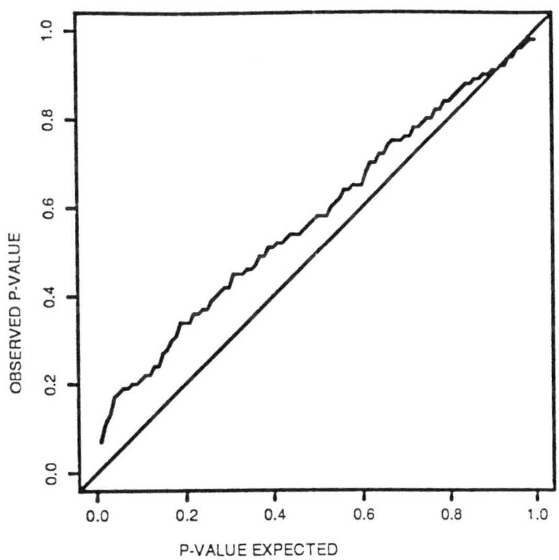

Figure 1b. *Q - Q plot of bootstrap P - values (Test for one mode, observations distributed according to a convolution of two uniforms)*

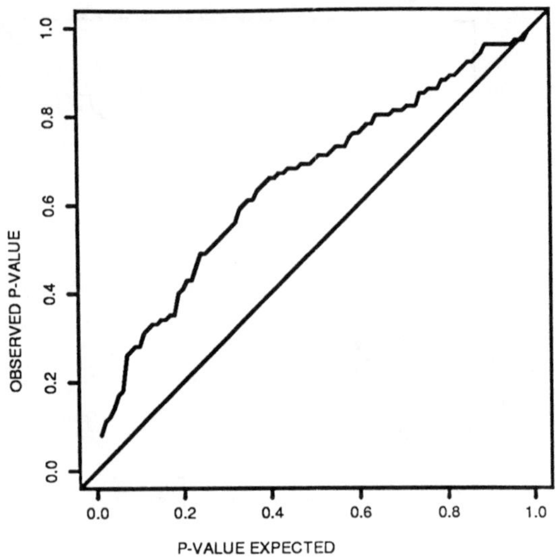

Figure 1c. *Q - Q plot of bootstrap P - values (Test for one mode, observations distributed according to a convolution of two uniforms)*

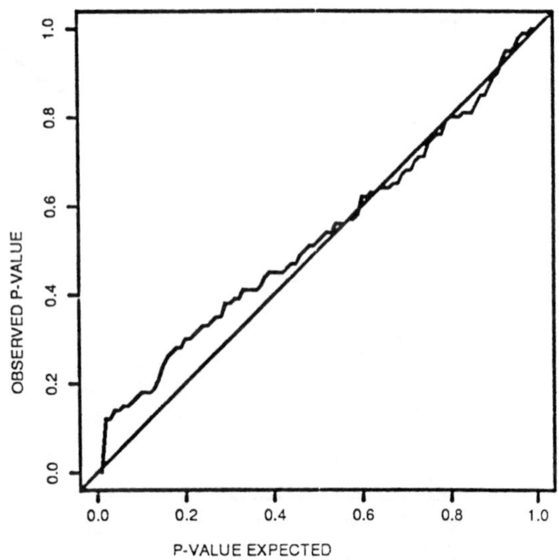

Figure 2a. *Q - Q plot of bootstrap P - values (Test for two modes, observations distributed according to a convolution of two uniforms)*

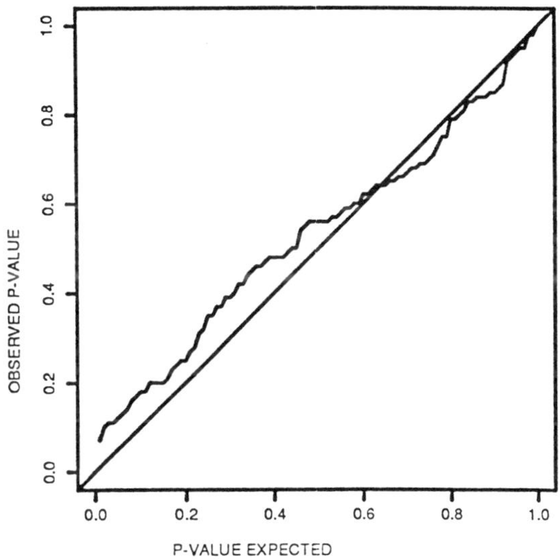

Figure 2b. *Q - Q plot of bootstrap P - values (Test for two modes, observations distributed according to a convolution of two uniforms)*

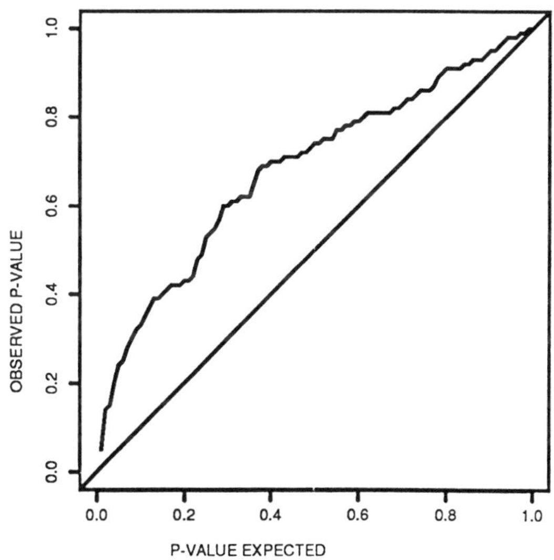

Figure 2c. *Q - Q plot of bootstrap P - values (Test for two modes, observations distributed according to a convolution of two uniforms)*

Chapter 5. Higher - order accuracy of bootstrap for smooth functionals.

1. Introduction. We come now back to bootstrap of smooth statistical functionals. In Chapter 1 we have studied under which conditions bootstrap of smooth functionals works. In this chapter we give a simple proof for the higher order accuracy of the bootstrap estimate for smooth functionals T. Under the assumption that T admits a higher order expansion we show that the expectation of a smooth function H of the studentized functional is estimated to order $O_P(n^{-1})$. For symmetric smooth functions H, the rate of convergence is $O_P(n^{-3/2})$. These results are in agreement with Hall (1986a, 1988, 1992), where these rates of convergence have been shown for functionals which are implicit or explicit functions of vector means (for a discussion of these rates see also Beran, 1987). The idea of the proof in this chapter is to make a direct comparison of the bootstrap estimate and the true distribution using the idea of Lindeberg's proof of the central limit theorem. The proof does not use Edgeworth expansions. The results of this chapter are also contained in Mammen(1990).

The approach in this chapter does not cover the case of indicator functions H. In particular, the important question is left open how accurately bootstrap estimates the distribution function of the (studentized) functional. Furthermore the results presented below do not imply rates for the coverage accuracy of confidence intervals. Nevertheless this approach offers a simple and intuitive way for understanding why bootstrap works for smooth functionals and why studentization may improve the accuracy of bootstrap. Furthermore the methods of this chapter may be used as first step for proving forthgoing results : The classical method for proving Edgeworth expansions of distribution functions is based on two steps. The first step is to show the convergence for expectations of smooth functions $H_t(x) = e^{itx}$ (i.e. characteristic functions). In the second step smoothing techniques are used. The major technical problem usually lies in the second step where stronger assumptions on the underlying distributions are needed (for instance: Cramér's condition in the case of linear functionals). For a new mathematical approach to carry out the second step see also Jeganathan (1990). The validity of Edgeworth expansions for higher - order von Mises functionals has been considered in Götze (1984) (see also Götze (1985, 1989)). For bootstrap samples the second step has been elaborated in a number of papers for functionals which are implicit or explicit functions of vector means. The unstudentized functional has been considered in Singh (1981). An $o_P(n^{-1/2})$ - rate of convergence has been proved in Babu and Singh (1984) for studentized functionals. The most general results for functions of vector means can be found in Hall (1986a, 1988, 1992) where exact rates are given for nonstudentized and studentized functionals, and where coverage accuracy and length of bootstrap confidence intervals are discusssed. In Abramovitch and Singh (1985) rates are given for bootstrap of Edgeworth corrected functionals. Bhattacharya (1987) compares bootstrap with higher order empirical Edgeworth expansions. Bhattacharya and Qumsiyeh (1989) show that bootstrap of the studentized statistic outperforms the two term empirical Edgeworth expansion in case of bowl shaped loss functions. Functionals which admit a quadratic approximation are studied in Beran (1982,1984). In particular, for studentized U - statistics rates are given for the bootstrap estimate of the expectation of a smooth function H. The bootstrap estimate of the distribution function of studentized U - statistics has been considered in Helmers (1991a), where an $o_P(n^{-1/2})$ rate has been shown for the bootstrap estimate.

Iteration of the bootstrap algorithm (bootstrapping the bootstrap estimate) has been proposed for improving the accuracy of bootstrap (see for instance Beran (1987), Hall and Martin (1988b), and Martin (1990). A rigid mathematical treatment of the higher order properties of iterated bootstrap procedures is rather complicated. Because of its simplicity we conjecture that the approach of this chapter can be used there.

The results of this chapter do not cover functionals which have no higher order smoothness. A statistical important example is the quantile functional. Bootstrap of quantiles is discussed for instance in Hall and Martin (1988a), Hall (1992), Falk and Reiss (1989 a,b), Falk and Kaufmann (1991), and Janas (1991).

2. Bootstrapping smooth functionals. We consider now again the bootstrap procedure of Chapter 1. Given a sample $X = (X_1, \dots, X_n)$ of n i. i. d. variables with unknown distribution P, this bootstrap procedure provides an estimate for $P(T(\widehat{P}_n) - T(P) \le t)$, where $\widehat{P}_n$ is the empirical distribution based on the sample. The bootstrap estimate is $P^*(T(\widehat{P}_n^*) - T(\widehat{P}_n) \le t)$. Here P^* denotes the conditional law $P(\dots|X)$. $\widehat{P}_n^*$ is the empirical distribution of a bootstrap sample $X^* = (X_1^*, \dots, X_n^*)$; that is, given X, the bootstrap sample X^* is an i.i.d. sample with distribution $\widehat{P}_n$.

We impose on T the following smoothness condition $\mathrm{DIFF}_T(k, C)$, $k \geq 1$, $C \geq 2$. In formulating it we use the following notation of Pfanzagl and Wefelmeyer (1985): given random variables R_n and constants $\delta_n \downarrow 0$ we write $R_n = \varepsilon_P(\delta_n)$ if $P(|R_n| \geq \varepsilon) = o(\delta_n)$ holds for every $\varepsilon > 0$. Moreover, it is allowed that the functional T may depend on the sample size n. This will not be indicated in the notation.

DIFF_T (k, C). *For* $1 \leq j \leq k$ *there exist functions* $t_{j,n} : \mathbf{R}^j \to \mathbf{R}$ *such that:*

(i) *For* $Q = \widehat{P}_n$ and $Q = \widehat{P}_n^*$ *the following expansion holds:*

$$T(Q) = T(P) + \tau_{1,n}(Q) + \dots + \tau_{k,n}(Q) + n^{-k/2}\, \varepsilon_P\left(n^{(1-k)/2}\right) \tag{2.1}$$

where

$$\tau_{j,n}(Q) = \int t_{j,n}(u_1, \dots, u_j)\,(Q - P)(du_1) \cdot \dots \cdot (Q - P)(du_j) \qquad (1 \leq j \leq k).$$

We choose the functions $t_{j,n}$ *such that* $t_{j,n}$ *is symmetric in its arguments and such that*

$$E(t_{j,n}(X_1, \dots, X_j) \mid X_1, \dots, X_{j-1}) = 0.$$

(ii) *The first term* $\tau_{1,n}(\widehat{P}_n)$ *is asymptotically not negligible :*

$$\inf_{n\geq 1} E\, t_{1,n}(X_1)^2 > 0\ .$$

(iii)
$$\sup_{n\geq 1} E\, | t_{j,n}(X_{i_1}, \dots, X_{i_j}) |^C < +\infty$$

for $1 \leq j \leq k$ and $1 \leq i_1, \dots, i_j \leq n$.

In (i) without loss of generality the functions $t_{j,n}$ can be chosen such that they are symmetric in their arguments because, for instance, for $j = 2$ one may define $t_{j,n}(u,v)$ as $1/2\ (t_{j,n}(u,v) + t_{j,n}(v,u))$. Moreover one gets $E(t_{j,n}(X_1, \dots, X_j) \mid X_1, \dots, X_{j-1}) = 0$ if, for instance, for $j = 2$ one chooses $t_{j,n}(u,v)$ as $t_{j,n}(u,v) - E(t_{j,n}(X_1,v)) - E(t_{j,n}(u,X_2))+$

$E(t_{j,n}(X_1,X_2))$. In (iii) double indices are used because, for instance, we need also the condition $\sup_{n\geq 1} E \mid t_{2,n}(X_1, X_1) \mid^C < +\infty$).

Similar differentiability conditions for functionals have been discussed in Reeds (1976). The function $t_{1,n}$ is also called the influence function of the functional T. The expansion (2.1) has been used in Pfanzagl and Wefelmeyer (1985) for k = 2. Bootstrap of functionals which admit a quadratic expansion has been considered in Beran (1982, 1984). The expansion (2.1) is related to Hoeffding's decomposition (see Chapter 1, Section 3). Condition DIFF_T (k, C) is discussed in Mammen (1990) for M - and L - functionals.

Often bootstrap works better by estimating the distribution of the studentized functional (see for instance Beran(1987) and the simulations in Chapter 1)

$$P\left(\sqrt{n}\,\frac{T(\hat{P}_n)-T(P)}{S_n(\hat{P}_n)} \leq t\right)$$

where $S_n^2(\hat{P}_n)$ is a variance estimate of $\sqrt{n}\ T(\hat{P}_n)$. We shall consider variance functionals $S_n(Q)$ which are defined for $Q = \hat{P}_n$, $Q = \hat{P}_n^*$, and Q = P and which fulfill the following smoothness condition $\mathrm{DIFF}_S(k, C)$:

DIFF_S(k, C). *For* $1 \leq j \leq k$ *there exist functions* $s_{j,n}$: $\mathbf{R}^j \rightarrow \mathbf{R}$ *such that:*

(i) *For* $Q = \hat{P}_n$ and $Q = \hat{P}_n^*$ *the following expansion holds:*

$$S_n^2(Q) = S_n^2(P) + \sigma_{1,n}(Q) + \ldots + \sigma_{k,n}(Q) + n^{-k/2}\ \varepsilon_P\left(n^{(1-k)/2}\right) \tag{2.2}$$

where

$$\sigma_{j,n}(Q) = \int s_{j,n}(u_1, \ldots, u_j)\ (Q-P)(du_1)\cdot \cdots \cdot (Q-P)(du_j) \qquad (1 \leq j \leq k).$$

As we have done in assumption DIFF_T(k, C) *we choose the functions* $s_{j,n}$ *such that* $s_{j,n}$

is symmetric in its arguments and such that

$$E(s_{j,n}(X_1, \dots, X_j) \mid X_1, \dots, X_{j-1}) = 0.$$

(ii) *Condition* DIFF$_T$(1, C') *holds for a* C' ≥ 2 *and*

$$(2.3) \qquad S_n^2(P) - \int t_{1,n}^2(u)\, P(du) \;=\; S_n^2(P) - n\, \mathrm{var}(\tau_{1,n}(\widehat{P}_n)) = O(n^{-1}).$$

(iii) $\qquad \sup_{n\geq 1} E \mid s_{j,n}(X_{i_1}, \dots, X_{i_j}) \mid^C < +\infty$

for $1 \leq j \leq k$ *and* $1 \leq i_1, \dots, i_j \leq n$.

Note that condition DIFF$_T$(k, C) does not imply that the variance of $n^{1/2}\,(T(\widehat{P}_n) - T(P))$ is finite. Therefore in (2.3) we compare the variance functional $S_n^2(P)$ with the variance of the linear approximation of $T(\widehat{P}_n)$. The $O(n^{-1})$ bound in condition (2.3) makes sense because of

$$n\, \mathrm{var}(\tau_{1,n}(\widehat{P}_n)) = n\, \mathrm{var}(\tau_{1,n}(\widehat{P}_n) + \dots + \tau_{k,n}(\widehat{P}_n)) + O(n^{-1}).$$

In particular, we consider the bootstrap variance estimator $S_n^2(\widehat{P}_n) = n\, \widehat{\mathrm{Var}}_{BOOT}$ and the jackknife variance estimator $S_n^2(\widehat{P}_n) = n\, \widehat{\mathrm{Var}}_{JACK}$:

$$\widehat{\mathrm{Var}}_{BOOT} = \mathrm{var}^*(T(\widehat{P}_n^*))$$

and

$$\widehat{\mathrm{Var}}_{JACK} = \frac{n-1}{n} \sum_{i=1}^n \Big(T(\widehat{P}_{n,(i)}) - \frac{1}{n} \sum_{j=1}^n T(\widehat{P}_{n,(j)}) \Big)^2.$$

Here for a random variable Z the quantity $\mathrm{var}^*(Z)$ denotes the conditional variance $E^*(Z - E^*(Z))^2$. $E^*(Z)$ is the conditional expectation $E(Z \mid X_1, \dots, X_n)$. $\widehat{P}_{n,(i)}$ are the empirical distributions of subsamples of size n - 1, i.e. $\widehat{P}_{n,(i)} = (n-1)^{-1} \sum_{j\neq i} \delta_{X_j}$, where δ_x denotes point mass in x. It has been shown in Mammen (1990) that condition

$DIFF_S(k, C)$ is fulfilled for the bootstrap and jackknife variance estimator for k = 2 and k = 3 under smoothness conditions on T (higher order Frechét differentiability). Furthermore, there for the bootstrap estimate the following stochastic expansion has been proved:

$$n\,\widehat{Var}_{BOOT} = n\ var^*(\tau_{1,n}(\widehat{P}_n^*) + \tau_{2,n}(\widehat{P}_n^*)) + n^{-3/2}\,\varepsilon_P(n^{-3/2}).$$

The first term of the right hand side of this equation is the (theoretical) bootstrap variance estimate of the quadratic approximation of $n^{1/2}\,T(\widehat{P}_n)$.

For $k \le 3$ this expansion implies the following property of the bootstrap variance estimate $S_n^2(\widehat{P}_n)$ (defining $S_n^2(P)$ as n var $\left(\tau_{1,n}(\widehat{P}_n) + \tau_{2,n}(\widehat{P}_n)\right)$)

VAR(k). *The follwing expansion holds*

$$(2.4)\qquad \left\{ S_n^2(\widehat{P}_n) - S_n^2(P) \right\}$$

$$- \left\{ n\ var^*\left(\tau_{1,n}(\widehat{P}_n^*) + \tau_{2,n}(\widehat{P}_n^*)\right) - n\ var\left(\tau_{1,n}(\widehat{P}_n) + \tau_{2,n}(\widehat{P}_n)\right) \right\} = O_P(n^{-k/2}).$$

We will consider only variance estimates which are asymptotically higher - order equivalent to the bootstrap variance estimate in the sense of (2.4) (for k = 2 or k = 3, respectively). Under smoothness conditions on T this includes the jackknife variance estimator $\widehat{Var}_{JACK}$. For higher - order Frechét differentiable functionals T this can be seen along the lines of the proof of Theorem 2 in Beran (1984) (see Mammen, 1990). A nice heuristic explanation for the higher - order equivalence of the bootstrap and the jackknife variance estimator has been given in Efron and Stein (1981) and Efron (1982), where it has been pointed out that the jackknife variance estimate can be interpreted as a bootstrap estimate of the variance of an 'estimated' linear approximation of the statistic T (see also equation (3.44) in Beran (1984)). In the proof of our theorem (see (4.22)) we will show that

$$(2.5)\qquad \left\{ n\ var^*\left(\tau_{1,n}(\widehat{P}_n^*) + \tau_{2,n}(\widehat{P}_n^*)\right) - n\ var\left(\tau_{1,n}(\widehat{P}_n) + \tau_{2,n}(\widehat{P}_n)\right) \right\}$$

$$- \left\{ n \operatorname{var}^* \left(\tau^*_{1,n}\left(\widehat{P}^*_n\right) \right) - n \operatorname{var} \left(\tau_{1,n}\left(\widehat{P}_n\right) \right) \right\} = O_P(n^{-3/2}),$$

where $T(\widehat{P}_n) + \tau^*_{1,n}(Q)$ is a linear approximation - depending on $\widehat{P}_n$ - of T(Q) for Q near to $\widehat{P}_n$ (see (4.22)). This explains why (2.4) holds for the jackknife variance estimate under appropriate regularity conditions. First order properties of the jackknife variance estimate have been considered also in Parr(1985), Sen(1988a,b), and Shao and Wu (1989). For a discussion of the bootstrap variance estimate see also Gill (1989) and Shao (1990).

We are now ready to state the main result of this chapter.

THEOREM. *For* m *large enough the following holds for functions* H *with* m *bounded derivatives:*

(i) *If* $DIFF_T(2, C)$ *holds for* C *large enough then*

$$\text{(2.6)} \qquad E^* H\left(\sqrt{n}\left(T\left(\widehat{P}^*_n\right) - T\left(\widehat{P}_n\right)\right)\right) - E\, H\left(\sqrt{n}\left(T\left(\widehat{P}_n\right) - T(P)\right)\right) = O_p\left(n^{-1/2}\right).$$

(ii) *If* $DIFF_T(3, C)$, $DIFF_S(2, C)$, *and* VAR(2) *hold for* C *large enough then*

$$\text{(2.7)} \qquad E^* H\left(\sqrt{n}\, \frac{T\left(\widehat{P}^*_n\right) - T\left(\widehat{P}_n\right)}{S\left(\widehat{P}^*_n\right)}\right) - E\, H\left(\sqrt{n}\, \frac{T\left(\widehat{P}_n\right) - T(P)}{S\left(\widehat{P}_n\right)}\right) = O_p(n^{-1}).$$

(iii) *If* $DIFF_T(4, C)$, $DIFF_S(3, C)$, *and* VAR(3) *hold for* C *large enough and* H *is a symmetric function:* H(x) = H(−x) *then*

$$\text{(2.8)} \qquad E^* H\left(\sqrt{n}\, \frac{T\left(\widehat{P}^*_n\right) - T\left(\widehat{P}_n\right)}{S\left(\widehat{P}^*_n\right)}\right) - E\, H\left(\sqrt{n}\, \frac{T\left(\widehat{P}_n\right) - T(P)}{S\left(\widehat{P}_n\right)}\right) = O_p(n^{-3/2}).$$

We think that with our approach one can show under stronger differentiability conditions

$$(2.9) \qquad E^* H\left(\sqrt{n}\, \frac{T(\hat{P}_n^*) - T(\hat{P}_n)}{S(\hat{P}_n^*)}\right) - E\, H\left(\sqrt{n}\, \frac{T(\hat{P}_n) - T(P)}{S(\hat{P}_n)}\right) = Z_n + O_p(n^{-2})$$

where Z_n is a random variable with $E\, Z_n = 0$ and $Z_n = O_p(n^{-3/2})$.

Under additional assumptions on the functions $t_{1,n}$ and $t_{2,n}$ formulas (2.6) and (2.7) can be also shown using Edgeworth expansions as in Beran (1982, Sect. 3.1). It is to be expected that under smoothness conditions on P and $t_{j,n}$, $s_{j,n}$ the theorem holds for nonsmooth functions H (for instance for indicator functions of intervals). Formulas (2.7 - 2.9) suggest that one - sided bootstrap confidence intervals have a coverage accuracy of order $O(n^{-1})$ while two-sided confidence intervals have a coverage accuracy of order $O(n^{-2})$. These rates of convergence have been proved by Hall (1986a, 1988, 1992) for the general class of functionals which are explicit or implicit functions of vector means. Note also that if one would prove (2.7) and (2.9) for indicator functions H the error terms $O_p(n^{-1})$ and $O_p(n^{-2})$ would not be sufficient for the calculation of coverage probabilities of bootstrap confidence intervals because it is not in general true that $P(X_n \leq x + O_p(n^{-\alpha})) = P(X_n \leq x) + O(n^{-\alpha})$ (for $\alpha > 0$). (Consider for instance a sequence of random variables Z_n with $P(Z_n = 1) = 1 - P(Z_n = 0) = n^{-\alpha+\varepsilon}$ for an ε with $0 < \varepsilon < \alpha$. Then for $x = 1/2$ and $X_n \equiv 0$ one gets $X_n - Z_n = O_p(n^{-\alpha})$, but $P(X_n \leq x) - P(Z_n \leq x) = n^{-\alpha+\varepsilon}$). For smooth H one can replace the error term $O_p(n^{-1})$ in (2.7) by $n^{-1/2}\, \varepsilon_p(n^{-1/2})$ (and $O_p(n^{-3/2})$ in (2.8) by $n^{-1}\, \varepsilon_p(n^{-1})$). This would show coverage errors of the bootstrap confidence intervals of order $o_p(n^{-1/2})$ and $o_p(n^{-1})$ (respectively) if one could show these expansions also for indicator functions H. For other bootstrap confidence intervals with higher order accuracy see Efron (1987), Diciccio and Romano (1988b), and Diciccio and Tibshirani (1987).

3. Some more simulations. Bootstrapping an M - estimate. How studentization improves the accuracy of bootstrap can also be seen in Figure 1 where simulation results are plotted for bootstrap of an M - estimate. Here, for a sample of n i. i. d. observations $X_1, \ldots, X_n$ we consider the following joint M - estimate $(\hat{T}, \hat{S})$ of location and scale:

$$\sum_{i=1}^{n} \psi\left(\frac{X_i - \widehat{T}}{\widehat{S}}\right) = 0,$$

$$\sum_{i=1}^{n} \psi^2\left(\frac{X_i - \widehat{T}}{\widehat{S}}\right) = (n - 1)\, \beta,$$

where β is the expectation of $\psi^2(U)$ for a standard normal variable U. ψ is the Huber M - function

$$\psi(x) = \begin{cases} -k & x \le -k \\ x & |x| < k \\ k & x \ge k. \end{cases}$$

k has been chosen as 1.14. This corresponds to minimax estimation in 10% contamination neighborhood of a normal distribution (see Huber, 1981). In our simulations we have data sets of 20 i. i. d. observations which have the normal mixture distribution of Model A in Chapter 1 (i.e. $.5\, N(.5, (.5)^2) + .5\, N(-.5, (.7)^2)$). As in Chapter 1 first we have run 16 000 simulations to determine u's and v's such that $P(\widehat{T} - T(P) \le u)$ and $P(\widehat{S}^{-1}(\widehat{T} - T(P)) \le v) = .01$, j/20 (for j =1, ..., 19), and .99 (where T(P) is the solution of $E_P\, \psi(x - T(P)) = 0$). Then we have generated 10 000 data sets and for every data set we have carried out the bootstrap procedure with 1000 replications. In the bootstrap resampling we have drawn 20 bootstrap observations $X_1^*, \ldots, X_{20}^*$ out of the set $\{ X_1 - \widehat{T}, \ldots, X_n - \widehat{T} \}$ and then used these bootstrap observations in (3.1) and (3.2) for calculating $(\widehat{T}^*, \widehat{S}^*)$. The bootstrap estimates of $P(\widehat{T} - T(P) \le u)$ and $P(\widehat{S}^{-1}(\widehat{T} - T(P)) \le v)$ are $P^*(\widehat{T}^* \le u)$ or $P^*(\widehat{T}^* / \widehat{S}^* \le v)$, respectively. Note that this bootstrap procedure differs slightly from the one described in the last section. The standard deviations of the bootstrap estimates are plotted in Figure 1. The bias of both bootstrap procedures is of lower order.

4. Proof of the theorem. We prove only (2.8). The proof goes as follows. First we derive a stochastic expansion of $\sqrt{n}\,(T(\widehat{P}_n) - T(P))\, S(\widehat{P}_n)^{-1} = \Delta_1 + \Delta_2 + \ldots$ (see (4.5)), where Δ_1 is a linear term, Δ_2 is a quadratic term, ... Using a Taylor expansion of $H(\sqrt{n}(T(\widehat{P}_n) - T(P))\, S(\widehat{P}_n)^{-1})$ around the linear term Δ_1 we get an expansion

$E\,H(\sqrt{n}(T(\widehat{P}_n) - T(P))\,S(\widehat{P}_n)^{-1}) = E\,H(\Delta_1)$ + higher order terms (see (4.7)) (STEP 1). In the next step by the same methods we get an expansion for the bootstrap estimate $E^*H(\sqrt{n}\,(T(\widehat{P}^*_n) - T(\widehat{P}_n))\,S(\widehat{P}^*_n)^{-1}) = E^*H(\Delta^*_1)$ + higher order terms (see (4.11)) (STEP 2). For the proof of (2.8) we have to compare the linear terms $E\,H(\Delta_1)$ and $E^*H(\Delta^*_1)$ and the higher order terms of these expansions. For the comparison of the linear terms we will use the idea of Lindeberg's proof of the central limit theorem (STEP 3). In STEP 4 the higher order terms are compared.

For simplicity we do not indicate each dependence on n in the notation. We write $\tau_j = \tau_{j,n}$, $\sigma_j = \sigma_{j,n}$, $t_j = t_{j,n}$, and $s_j = s_{j,n}$. Furthermore for simplicity we assume that

$$t_2(u,v) = 0 \qquad \text{for } u = v.$$

This implies that τ_2 is purely quadratic, i. e. that the diagonal terms of τ_2 vanish. One can show that this is no real restriction. In particular one may choose $t_1(u)$ as $t_1(u) + n^{-1}\,(t_2(u,u) - E_P(t_2(X_1,X_1)))$. Note that

$$(4.1) \qquad E\,n^{-3/2}\sum_{i=1}^{n} t_2(X_i,X_i) = E^*\,n^{-3/2}\sum_{i=1}^{n} t_2(X^*_i,X^*_i) + O_P(n^{-1})$$

and

$$(4.2) \qquad E\,H'\left(\sqrt{n}\,\frac{T(\widehat{P}_n) - T(P)}{S(\widehat{P}_n)}\right) = O(n^{-1/2}).$$

To prove (4.2) one shows first $E\,H'(\sqrt{n}(T(\widehat{P}_n) - T(P))\,S(\widehat{P}_n)^{-1}) - E\,H'(\sqrt{n}\;\tau_1(\widehat{P}_n)\,S_n^{-1}(P)) = O(n^{-1/2})$. Furthermore by an application of the Berry - Esséen inequality for the linear approximation $\sqrt{n}\;\tau_1(\widehat{P}_n)\,S_n^{-1}(P)$ one gets $E\,H'(\sqrt{n}\;\tau_1(\widehat{P}_n)\,S_n^{-1}(P)) - E\,H'(Z) = O(n^{-1/2})$, where Z is a random variable with standard normal distribution. This shows

$$(4.3) \qquad E\,H'\left(\sqrt{n}\,\frac{T(\widehat{P}_n) - T(P)}{S(\widehat{P}_n)}\right) - E\,H'(Z) = O(n^{-1/2}).$$

Now the symmetry of H implies $E\,H'(Z) = 0$. This shows (4.2).

STEP 1. ASYMPTOTIC EXPANSION OF E $H(\sqrt{n}(T(\hat{P}_n) - T(P))\, S(\hat{P}_n)^{-1})$: First we state a stochastic expansion for $S(\hat{P}_n)^{-1}$. Note that for a constant C' condition $DIFF_S(k, C)$ implies (if C is chosen large enough):

For $1 \le j \le k$ there exist functions $r_j = r_{j,n}: \mathbf{R}^j \to \mathbf{R}$ such that

(i) For $Q = \hat{P}_n$ and $Q = \hat{P}_n^*$ the following expansion holds:

$$S_n^{-1}(Q) = S_n^{-1}(P) + \rho_1(Q) + \ldots + \rho_k(Q) + n^{-k/2}\, \varepsilon_p\left(n^{(1-k)/2}\right) \tag{4.4}$$

where

$$\rho_j(Q) = \int r_j(u_1, \ldots, u_j)\ (Q - P)(du_1) \cdot\ldots\ (Q - P)(du_j) \qquad (1 \le j \le k).$$

(ii) r_j is symmetric in its arguments and

$$E(r_j(X_1, \ldots, X_j) \mid X_1, \ldots, X_{j-1}) = 0 .$$

(iii) $\quad \sup_{n \ge 1} E \mid r_j(X_{i_1}, \ldots, X_{i_j}) \mid^{C'} < +\infty$ for $1 \le j \le k$ and $1 \le i_1, \ldots, i_j \le n$.

With the expansion (4.4) for $Q = \hat{P}_n$ and k =3 we get

$$\sqrt{n}\ \frac{T(\hat{P}_n) - T(P)}{S_n(\hat{P}_n)} = \Delta_1 + \ldots + \Delta_4 + n^{-3/2}\ \varepsilon_p(n^{-3/2}), \tag{4.5}$$

where

$$\Delta_1 = \sqrt{n}\ \tau_1(\hat{P}_n)\ S_n^{-1}(P),$$

$$\Delta_2 = \sqrt{n}\ \tau_1(\hat{P}_n)\ \rho_1(\hat{P}_n) + \sqrt{n}\ \tau_2(\hat{P}_n)\ S_n^{-1}(P),$$

$$\Delta_3 = \sqrt{n}\ \tau_1(\hat{P}_n)\ \rho_2(\hat{P}_n) + \sqrt{n}\ \tau_2(\hat{P}_n)\ \rho_1(\hat{P}_n) + \sqrt{n}\ \tau_3(\hat{P}_n)\ S_n^{-1}(P),$$

$$\Delta_4 = \sqrt{n}\ \tau_1(\widehat{P}_n)\ \rho_3(\widehat{P}_n) + \ldots + \sqrt{n}\ \tau_4(\widehat{P}_n)\ S_n^{-1}(P).$$

Since $r_j(X_{i_1}, \ldots, X_{i_j})$ and $t_j(X_{i_1}, \ldots, X_{i_j})$ have uniformly bounded fourth moments one can show for $1 < j \leq 4$:

$$E\,\Delta_j^2 = O(n^{1-j}). \tag{4.6}$$

(4.5) and (4.6) imply

$$E\ H\left(\sqrt{n}\ \frac{T(\widehat{P}_n) - T(P)}{S_n(\widehat{P}_n)}\right) \tag{4.7}$$

$$= E\ H(\Delta_1 + \Delta_2 + \Delta_3) + O(n^{-3/2})$$

$$= E\ H(\Delta_1) + E\,H'(\Delta_1)(\Delta_2 + \Delta_3)$$

$$+ \frac{1}{2}\ E\,H''(\Delta_1)\ \Delta_2^{\,2} + O(n^{-3/2}).$$

Here in the second equality we have used the fact that $|H^{(k)}|$ is bounded for $k \leq 3$. For the first equality note that because of the definition of $\varepsilon_p(n^{-3/2})$ there exists a random variable $Z = Z_n$ with $|Z| \leq 1$ such that $E\,|\,H(\Delta_1 + \ldots + \Delta_4 + n^{-3/2}\ \varepsilon_p(n^{-3/2})) - H(\Delta_1 + \ldots + \Delta_3)| \leq E\,|\,H(\Delta_1 + \ldots + \Delta_4 + n^{-3/2} Z) - H(\Delta_1 + \ldots + \Delta_3)| + 2\sup_x |\,H(x)\,|\,o(n^{-3/2}) \leq \sup_x |\,H'(x)\,|\,E\,|\,\Delta_4 + n^{-3/2} Z\,| + o(n^{-3/2}) = O(n^{-3/2})$.

STEP 2. ASYMPTOTIC EXPANSION OF THE BOOTSTRAP ESTIMATE $E^*(\sqrt{n}\,(T(\widehat{P}_n^*) - T(\widehat{P}_n))\,S(\widehat{P}_n^*)^{-1})$: The bootstrap counterpart of the expansion (4.7) can be derived as follows. First we define modifications t_1^* and t_2^* of t_1 and t_2 which fulfill $E^* t_1^*(X_i^*) = 0$ and $E^*\left(t_2^*(X_i^*, X_j^*)\,|\,X_i^*\right) = 0$. This can be done by setting

$$t_1^*(u) = [t_1(u) - 1/n \sum_{i=1}^{n} t_1(X_i)] + 2\,(n-1)\,n^{-1}\,t_2^*(u),$$

$$t_2^*(u,v) = \begin{cases} [t_2(u,v) - t_2^*(u) - t_2^*(v) - t_0^*], & \text{if } u \neq v, \\ 0, & \text{if } u = v, \end{cases}$$

$$t_2^*(u) = 1/n \sum_{i=1}^{n} t_2(u,X_i) - t_0^*,$$

$$t_0^* = 1/n^2 \sum_{i,j=1}^{n} t_2(X_i\,,\,X_j\,).$$

Define now $\quad \tau_0^* = -\, n^{-1}\, t_0^*,$

$$\tau_1^* = n^{-1} \sum_{i=1}^{n} t_1^*(X_i^*)\,,$$

and

$$\tau_2^* = n^{-2} \sum_{i,j=1}^{n} t_2^*(X_i^*, X_j^*)\,.$$

Note that t_1^* and t_2^* are chosen such that $\tau_1(\widehat{P}_n^*) + \tau_2(\widehat{P}_n^*) - \tau_1(\widehat{P}_n) - \tau_2(\widehat{P}_n) = \tau_0^* + \tau_1^* + \tau_2^*$.

With this notation we get from $DIFF_T(4, C)$

$$\sqrt{n}\,\left(T(\widehat{P}_n^*) - T(\widehat{P}_n)\right) = \tau_0^* + \ldots + \tau_4^* + n^{-3/2}\, \varepsilon_P(\,n^{-3/2}\,),$$

where for $j = 3$ and $j = 4$

$$\tau_j^* = \tau_j(\widehat{P}_n^*) - \tau_j(\widehat{P}_n).$$

Furthermore using the expansion (1.2) of $S_n^{-1}(Q)$ for $Q = \widehat{P}_n$ and $Q = \widehat{P}_n^*$ with $k = 3$ we get

$$S_n^{-1}(\widehat{P}_n^*) = S_n^{-1}(\widehat{P}_n) + \rho_1^* + \ldots + \rho_3^* + n^{-3/2}\, \varepsilon_P(\,n^{-3/2}\,),$$

where for $1 \leq j \leq 3$

$$\rho_j^* = \rho_j(\hat{P}_n^*) - \rho_j(\hat{P}_n).$$

Now we get

$$(4.8) \qquad \sqrt{n}\ \frac{T(\hat{P}_n^*) - T(\hat{P}_n) - \tau_0^*}{S_n(\hat{P}_n^*)}$$

$$= \Delta_1^* + \Delta_2^* + \Delta_3^* + \Delta_4^* + n^{-3/2}\ \varepsilon_P(n^{-3/2}),$$

where

$$\Delta_1^* = \sqrt{n}\ \tau_1^*\ S_n^{-1}(\hat{P}_n),$$

$$\Delta_2^* = \sqrt{n}\ \tau_1^*\ \rho_1^* + \sqrt{n}\ \tau_2^*\ S_n^{-1}(\hat{P}_n),$$

$$\Delta_3^* = \sqrt{n}\ \tau_1^*\ \rho_2^* + \sqrt{n}\ \tau_2^*\ \rho_1^* + \sqrt{n}\ \tau_3^*\ S_n^{-1}(\hat{P}_n),$$

$$\Delta_4^* = \sqrt{n}\,\tau_1^*\ \rho_3^* + \ldots + \sqrt{n}\ \tau_4^*\ S_n^{-1}(\hat{P}_n).$$

Furthermore for $1 \leq j \leq 4$ one can show the analogue of (4.6)

$$(4.9) \qquad E\left(\Delta_j^*\right)^2 = O(n^{1-j}).$$

This implies

$$(4.10) \qquad E^*\left(\Delta_j^*\right)^2 = O_P(n^{1-j}).$$

As in (4.7) we deduce from (4.8), (4.10), and $\tau_0^* = O_P(n^{-2})$ that

$$(4.11) \qquad E^* H\left(\sqrt{n}\,\frac{T(\hat{P}_n^*) - T(\hat{P}_n)}{S_n(\hat{P}_n^*)}\right)$$

$$= E^* H\left(\sqrt{n}\,\frac{T(\hat{P}_n^*) - T(\hat{P}_n) - \tau_0^*}{S_n(\hat{P}_n^*)}\right) + O_P(n^{-3/2})$$

$$= E^* H(\overset{*}{\Delta}_1) + E^* H'(\overset{*}{\Delta}_1)\,(\overset{*}{\Delta}_2 + \overset{*}{\Delta}_3)$$

$$+ \frac{1}{2} E^* H''(\overset{*}{\Delta}_1)\left(\overset{*}{\Delta}_2\right)^2 + O_P(n^{-3/2}).$$

The proof of (2.8) now follows from a comparison of (4.7) and (4.11). Here we show only that

$$(4.12) \qquad E^* H(\overset{*}{\Delta}_1) - E\, H(\Delta_1) = O_P(n^{-3/2})$$

and

$$(4.13) \qquad \sqrt{n}\,(E^* H'(\overset{*}{\Delta}_1)\, \tau_2^*\, S_n^{-1}(\hat{P}_n)$$

$$- E\, H'(\Delta_1)\, \tau_2(\hat{P}_n)\, S_n^{-1}(P)) = O_P(n^{-3/2}) \; .$$

The comparison of the other terms of (4.7) and (4.11) follows analogously.

STEP 3 . COMPARISON OF THE LINEAR TERMS E $H(\Delta_1)$ AND $E^*H(\overset{*}{\Delta}_1)$ (PROOF OF (4.12)).

We use the idea of Lindeberg's proof of the central limit theorem. For notational convenience we want to use the same expectation operator for the original sample and for the bootstrap sample. Therefore we introduce random variables $\tilde{X}_i$ $(1 \le i \le n)$ which are conditionaly independent (given $X_1, \dots, X_n$) and which have conditional distribution $\mathcal{L}^*(\tilde{X}_i) = \mathcal{L}(\tilde{X}_i \mid X_1, \dots, X_n) = \mathcal{L}(X_i)$. Then one can write for instance $E\, t_1(X_i)$ -

$E^* \, t_1^*(X_i^*) = E^*[t_1(\widetilde{X}_i) - t_1^*(X_i^*)].$

Let $$Y_i = t_1(\widetilde{X}_i) \; S_n^{-1}(P),$$

$$Y_i^* = t_1^*(X_i^*) \; S_n^{-1}(\widehat{P}_n),$$

$$Z_i = 1/\sqrt{n} \sum_{j<i} Y_j^* + 1/\sqrt{n} \sum_{j>i} Y_j.$$

Then

$$E^* H(\Delta_1^*) - E \;\; H(\Delta_1)$$

$$= E^* \left[H(\, 1/\sqrt{n} \sum_{i=1}^n Y_i^* \,) - H(\, 1/\sqrt{n} \sum_{i=1}^n Y_i \,) \right]$$

$$= \sum_{i=1}^n E^* \; H(Z_i + Y_i^* / \sqrt{n}\,) - H\,(Z_i + Y_i/\sqrt{n}\,)$$

$$= S_1 + \ldots + S_4 + R$$

where

$$S_j = n^{-j/2} \; 1/\,j! \;\; E^* \sum_{i=1}^n H^{(j)}(Z_i) \; \left((Y_i^*)^j - Y_i^j \right),$$

$$R = n^{-5/2} \frac{1}{120} \;\; E^* \sum_{i=1}^n H^{(5)} (\widetilde{Z}_i) \;\; \left((Y_i^*)^5 - Y_i^5 \right)$$

for a random variable $\widetilde{Z}_i$ lying between $Z_i + Y_i / \sqrt{n}$ and $Z_i + Y_i^* / \sqrt{n}$.

Now $S_1 + \ldots + S_4 + R = O_P(n^{-3/2})$ follows from the boundedness of $H^{(j)}$ $(1 \le j \le 5)$ and

(4.14) $$E^* \; H^{(3)}(Z_i) = O_P(n^{-1/2}),$$

(4.15) $$E^* Y_i^* = E^* Y_i \ (=0),$$

(4.16) $$E^* (Y_i^*)^2 = E^* Y_i^2 + O_P(n^{-3/2}),$$

(4.17) $$E^* (Y_i^*)^j = E^* Y_i^j + O_P(n^{-1/2}) \qquad \text{(for } j = 3,\ 4),$$

(4.18) $$E^* (Y_i^*)^5 = O_P(1), \text{ and } E^* Y_i^5 = O_P(1).$$

We now show (4.14) and (4.16). The remaining approximations are verified easily. For the proof of (4.14) note first that the symmetry of H implies

(4.19) $$E^* H^{(3)}(Z_i) = E^* H^{(3)}(Z_i) - E\ H^{(3)}\left(1/\sqrt{n} \sum_{j=1}^{n-1} U_j\right)$$

for standard normal distributed independent random variables $U_1, \ldots, U_n$. Now the right hand side of this equation can be treated by Lindeberg's argument as above. For the proof of (4.16) note first that

(4.20) $$E^* \left(t_2^*(X_1^*, X_2^*)^2\right) - E\ \left(t_2(X_1, X_2)^2\right) = O_P(n^{-1/2}).$$

Then

(4.21) $$E^* (Y_i^*)^2 - E^* Y_i^2$$

$$= S_n^{-2}(\hat{P}_n)\, E^* \left(t_1^*(X_1^*)^2\right) - S_n^{-2}(P)\, E\left(t_1(X_1)^2\right)$$

$$= S_n^{-2}(\hat{P}_n)\left[E^* \left(t_1^*(X_1^*)^2\right) + \frac{n-1}{n^2}\left\{E^* \left(t_2^*(X_1^*, X_2^*)^2\right) - E\ \left(t_2(X_1, X_2)^2\right)\right\}\right]$$

$$- S_n^{-2}(P)\, E\left(t_1(X_1)^2\right) + O_P(n^{-3/2}) \qquad \text{(see (4.20))}$$

$$= S_n^{-2}(\hat{P}_n)\left[S_n^2(\hat{P}_n) - S_n^2(P) + E\ \left(t_1(X_1)^2\right)\right]$$

$$- S_n^{-2}(P)\,E\left(t_1(X_1)^2\right) + O_P(n^{-3/2}) \qquad \text{(see (2.4))}$$

$$= S_n^{-2}(\hat{P}_n)\left[S_n^2(\hat{P}_n) - S_n^2(P)\right]$$

$$+\left[S_n^{-2}(\hat{P}_n) - S_n^{-2}(P)\right]\,E\left(t_1(X_1)^2\right) + O_P(n^{-3/2})$$

$$= S_n^{-2}(\hat{P}_n)\left[S_n^2(\hat{P}_n) - S_n^2(P)\right]$$

$$+\left[S_n^{-2}(\hat{P}_n) - S_n^{-2}(P)\right]\,S_n^2(P) + O_P(n^{-3/2}) \qquad \text{(see (2.3))}$$

$$= O_P(n^{-3/2}).$$

Let us shortly mention here the proof of (2.5)

(4.22) $\left\{ S_n^2(\hat{P}_n) - S_n^2(P) \right\}$

$$= \left\{ n\,\mathrm{var}^*\left(\tau_1\left(\hat{P}_n^*\right) + \tau_2\left(\hat{P}_n^*\right)\right) - n\,\mathrm{var}\left(\tau_1\left(\hat{P}_n\right) + \tau_2\left(\hat{P}_n\right)\right) \right\} + O_P(n^{-3/2})$$

$$= \left[E^*\left(t_1^*(X_1^*)^2\right) - E\left(t_1(X_1)^2\right)\right] + (n-1)\,n^{-2}\left[E^*\left(t_2^*(X_1^*,X_2^*)^2\right) - E\left(t_2(X_1,X_2)^2\right)\right]$$

$$+ O_P(n^{-3/2})$$

$$= \left[E^*\left(t_1^*(X_1^*)^2\right) - E\left(t_1(X_1)^2\right)\right] + O_P(n^{-3/2}) \qquad \text{(see (4.16))}$$

$$= \left\{ n\,\mathrm{var}^*\left(\tau_1^*\left(\hat{P}_n^*\right)\right) - n\,\mathrm{var}\left(\tau_1\left(\hat{P}_n\right)\right) \right\} + O_P(n^{-3/2}).$$

STEP 4 . COMPARISON OF THE HIGHER ORDER TERMS OF E $H(\sqrt{n}(T(\hat{P}_n) - T(P))$ $S(\hat{P}_n)^{-1})$ AND $E^*(\sqrt{n}\,(T(\hat{P}_n^*) - T(\hat{P}_n))\,S(\hat{P}_n^*)^{-1})$ (PROOF OF (4.13)).

For simplicity we suppose $S_n(P) = 1$. First note that

(4.23) $$\sqrt{n}\, E\, H'(\Delta_1)\ \tau_2(\widehat{P}_n)\ S_n^{-1}(P)$$

$$= E \sum_{i \neq j} H'\left(1/\sqrt{n} \sum_{k=1}^{n} t_1(X_k)\right) n^{-3/2}\ t_2(X_i\,,X_j\,)$$

$$= E \sum_{i \neq j} H'\left(1/\sqrt{n} \sum_{k \neq i,j} t_1(X_k)\right) n^{-3/2}\ t_2(X_i,X_j)$$

$$+ \sum_{i \neq j} H''\left(1/\sqrt{n} \sum_{k \neq i,j} t_1(X_k)\right) n^{-5/2}\ t_2(X_i\,,X_j)\ (\,t_1(X_i) + t_1(X_j)\,) + \ldots$$

$$= \ E \sum_{i \neq j} H'''\left(1/\sqrt{n} \sum_{k \neq i,j} t_1(X_k)\right) n^{-5/2}\ t_2(X_i\,,X_j)\ t_1(X_i)\ t_1(X_j)$$

$$+ H^{(4)}\left(1/\sqrt{n} \sum_{k \neq i,j} t_1(X_k)\right) n^{-3}\ t_2(X_i\,,X_j)\ t_1^2(X_i)\ t_1(X_j)\ +\ O(n^{-3/2})$$

where $E(t_2(X_1, X_2) \mid X_2) = 0$ has been used. For the bootstrap expression we get

(4.24) $$\sqrt{n}\, E^*\, H'\left(\Delta_1^*\right)\ \tau_2^*\ S_n^{-1}(\widehat{P}_n)$$

$$= E^* \sum_{i \neq j} H'\left(\ S_n^{-1}(\widehat{P}_n)\ 1/\sqrt{n} \sum_{k=1}^{n} t_1^*(X_k^*)\ \right) n^{-3/2}\, S_n^{-1}(\widehat{P}_n)\, t_2^*\left(X_i^*,X_j^*\right).$$

The right hand side of this equation can now be expanded as in (4.23). To prove (4.13) we apply

$$E\ \ H'''\left(1/\sqrt{n} \sum_{k \neq i,j} t_1(X_k)\right) = O(n^{-1/2}),$$

$$E\ \ H'''\left(1/\sqrt{n} \sum_{k \neq i,j} t_1(X_k)\right) - E^*\ \ H'''\left(1/\sqrt{n} \sum_{k \neq i,j} t_1^*(X_k^*)\right) = O_P(n^{-1}),$$

$$E\ \ t_2(X_i,X_j)\ \ t_1^k(X_i)\ \ t_1(X_j)$$

$$- E^* \; t_2^*\left(X_i^*, X_j^*\right) \; (t_1^*(X_i^*))^k \; t_1^*(X_j^*) = O_P(n^{-1/2}) \text{ for } k = 1 \text{ and } 2.$$

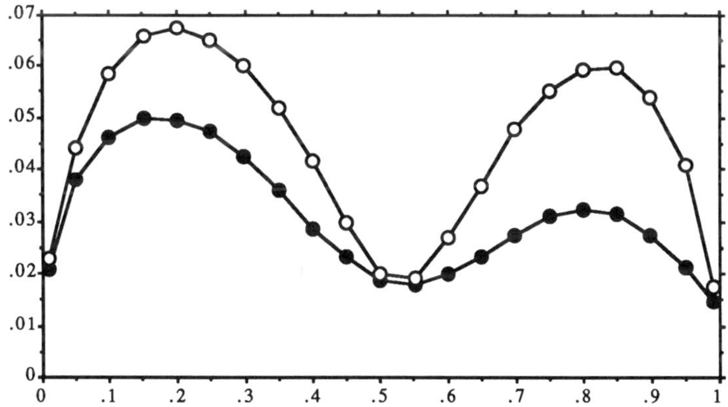

Figure 1. *Monte Carlo estimates of the* **standard deviation** *of bootstrap estimates for the distribution function of a unstudentized M estimate and a studentized M estimate, plotted against the distribution function. 20 i. i. d. observations with a normal mixture distribution.* ○ *bootstrap of the unstudentized estimate,* ● *bootstrap of the studentized estimate*

Chapter 6. Bootstrapping linear models.

1. Bootstrapping the least squares estimator. In the next both chapters we consider bootstrap of least squares estimates and M - estimates in linear models with nonrandom design and i.i.d. error variables. We show that bootstrap works under weaker conditions than asymptotic normality. This will be done for least squares estimates in this chapter and for M - estimates in the next chapter. This result has an important practical consequence. It shows that bootstrap can be applied in cases where the classical normal approximation breaks down. By contrast, in the last chapter we have shown higher - order accuracy of the bootstrap estimate. This result says only that bootstrap works better if the normal approximation is already quite good. In many applications this additional accuracy has less practical importance.

This chapter contains material from Bickel and Freedman (1983) and Mammen (1989b). We consider the linear model

$$Y_i = X_i^T \beta + \varepsilon_i \qquad (i = 1, \dots, n) .$$

where $\mathbf{Y} = (Y_1, \dots, Y_n)^T$ is the observation vector, $\mathbf{X} = (X_1, \dots, X_n)^T$ is a $n \times p$ (observed) design matrix, and $\varepsilon = (\varepsilon_1, \dots, \varepsilon_n)^T$ is the error vector consisting of n i.i.d. mean 0 variables with distribution P. In this chapter we study bootstrapping the distribution of the least squares estimator

$$\widehat{\beta} = (\mathbf{X}^T\mathbf{X})^{-1} \mathbf{X}^T\mathbf{Y}.$$

In Chapter 8 we will consider bootstrap in random design models. Then the bootstrap procedure is based on resampling the observations *and* the design vectors. We will see that for high dimensions this stochastic model is too complicated for the bootstrap and that bootstrap does not work in this model always satisfactorily. In the model with nonrandom design which we consider in this and in the next chapter one resamples only the one - dimensional observations. This is the reason why bootstrap works here so accurately (see also the simulations in the next chapter). Furthermore, the bootstrap procedure uses explicitly the knowledge of every design point whereas the normal approximation uses only the knowledge of the matrix $\mathbf{X}^T\mathbf{X} = \sum_{i=1}^n X_i X_i^T$. This will explain the superiority of bootstrap (in particular, for unbalanced designs as in the presence of leverage points). In Navidi (1989) the superiority of bootstrap has been shown using higher order Edgeworth expansions. In particular, for unbalanced designs bootstrap has a faster rate of convergence than the normal approximation. In contrat to the i. i. d. case (see the last chapter) this holds for the *unstudentized* (least squares) estimate.

In this and in the following chapters we use an asymptotic approach in which the dimension p of the linear model and of the design vectors $X_1, \ldots, X_n$ may depend on the sample size n. This will not be indicated in the notation. We will also allow the dimension p to grow with the sample size n. This feature is important because in many applications models are used where the dimension p is not small compared with n. An asymptotic approach in which p is fixed is misleading because the high dimensionality of the model is lost asymptotically. Asymptotics with increasing p have been also proposed in Huber (1981). The following theorem gives a necessary and sufficient condition for asymptotic normality of all linear contrasts.

THEOREM 1 (HUBER (1981)). *Suppose* $\sigma^2 = \mathrm{E}\,\varepsilon_i^2 < +\infty$ *and assume that* $\Sigma_{i=1}^n X_i X_i^T$ *is positively definite for all* n. *Then the following statements are equivalent.*

(i) $$(c_n^T (\mathbf{X}^T\mathbf{X})^{-1} c_n)^{-1/2}\, c_n^T(\hat\beta - \beta)$$

converges weakly to $N(0,\sigma^2)$ *for all sequences* c_n.

(ii) $$\max_{1\le i\le n} H_{ii} \to 0 \qquad or \qquad \mathcal{L}(\varepsilon_i) = N(0,\sigma^2)\,.$$

Here H_{ii} *is the* i *th diagonal element of the hat matrix* $H = \mathbf{X}(\mathbf{X}^T\mathbf{X})^{-1}\mathbf{X}^T$.

PROOF. For simplicity assume $\sum_{i=1}^n X_i X_i^T = \mathbf{X}^T\mathbf{X} = I$ (If this does not hold replace X_i by $(\mathbf{X}^T\mathbf{X})^{-1/2} X_i$ and the parameter β by $(\mathbf{X}^T\mathbf{X})^{1/2}\beta$. This corresponds to the usual $\sqrt{n}$ rescaling in a one dimensional location problem.). Then $H_{ii} = X_i^T X_i = \|X_i\|^2$ and $c_n^T(\hat\beta - \beta) = \sum_{i=1}^n (c_n^T X_i)\,\varepsilon_i$. Now "(ii) $\Rightarrow$ (i)" follows from the central limit theorem. For the proof of "(i) $\Rightarrow$ (ii)" assume that the error variables ε_i are not normally distributed and that (for instance) $H_{11} \to D > 0$. Put $c_n = X_1 / \|X_1\|$. Then a subsequence of $\mathcal{L}(\sum_{i=2}^{n} (c_n^T X_i)\, \varepsilon_i)$ converges weakly to a distribution μ, say. If (i) would hold this would imply that $N(0,\sigma^2) = \mathcal{L}(\sqrt{D}\, \varepsilon_1) * \mu$. According to a result of Levy (see, e.g., Feller (1966), p. 498) then $\mathcal{L}(\sqrt{D}\, \varepsilon_1)$ and μ must be normal distributions. Therefore $\mathcal{L}(\varepsilon_1) = N(0,\sigma^2)$.

We want now to study under which conditions bootstrap works. For this purpose define $\psi_{np}(Q)$ as the distribution of $\text{cov}(\hat{\beta})^{-1/2}(\hat{\beta} - \beta)$ if the errors are distributed according to the distribution Q. Furthermore put $\psi_{npc}(Q)$ as the distribution of $c^T(\hat{\beta} - \beta)$ under the error distribution Q. We want to study under which conditions the following (stochastic) approximations are accurate

$$\psi_{np}(\tilde{P}_n) \approx \psi_{np}(P),$$

and

$$\psi_{npc}(\tilde{P}_n) \approx \psi_{npc}(P)$$

where $\tilde{P}_n$ is an estimate of P. Because we have only assumed that P has *mean 0* an appropriate estimate $\tilde{P}_n$ for P is the *shifted empirical distribution* of the residuals $\hat{\varepsilon}_i = Y_i - X_i^T \hat{\beta}$, i.e. $\tilde{P}_n(B) = n^{-1} \sum_{i=1}^{n} I(\hat{\varepsilon}_i - n^{-1} \sum_{j=1}^{n} \hat{\varepsilon}_j \in B)$. We measure the distance between two distributions μ and ν by the Mallows distance

$$d_2(\mu,\nu) = \inf\{[E\,\|U - V\|^2]^{1/2} : \mathcal{L}(U) = \mu\,,\ \mathcal{L}(V) = \nu\}.$$

The next theorem describes the dependence of $\psi_{np}(Q)$ (and $\psi_{npc}(Q)$) on the underlying error distribution Q.

THEOREM 2 (BICKEL AND FREEDMAN (1983)). *Suppose that* $\mathbf{X}^T\mathbf{X}$ *is nonsingular. Then for two probability measures* P *and* Q *with mean* 0 *the following holds*

(i) $$d_2(\psi_{np}(P)\,,\ \psi_{np}(Q)) \le \sqrt{p}\; d_2(P,Q),$$

(ii) $$d_2(\psi_{npc}(P)\,,\,\psi_{npc}(Q)) \le d_2(P,Q),$$

$$\text{if } c^T(\mathbf{X}^T\mathbf{X})^{-1}c = 1.$$

PROOF. As in the proof of Theorem 1 suppose $\mathbf{X}^T\mathbf{X} = I$. Choose $(\xi_1,\eta_1),\ldots,(\xi_n,\eta_n)$ i.i.d. with marginal distributions $\mathcal{L}(\xi_1) = P$ and $\mathcal{L}(\eta_1) = Q$ and with $E(\xi_1 - \eta_1)^2 = d_2^2(P,Q)$.

Then

$$d_2^2(\psi_{np}(P)\,,\,\psi_{np}(Q)) \le E\,\| \textstyle\sum_{i=1}^n X_i(\xi_i - \eta_i)\|^2$$

$$= E \textstyle\sum_{i=1}^n \|X_i\|^2 (\xi_i - \eta_i)^2$$

$$= p\, d_2^2\,(P,Q),$$

because H is the projection onto a p dimensional space and therefore $\sum_{i=1}^n \|X_i\|^2 = \Sigma_{i=1}^n H_{ii} = \text{spur } H = p$.

(ii) can be shown similarly.

Theorem 2 can be applied to show consistency of bootstrap.

COROLLARY (BICKEL AND FREEDMAN (1983)). *Suppose that* $\mathbf{X}^T\mathbf{X}$ *is nonsingular,* $\sigma^2 = E\,\varepsilon_i^2 < +\infty$, and $c^T(\mathbf{X}^T\mathbf{X})^{-1}c = 1$ *for a sequence* $c = c_n$. *Then*

(i) $$d_2(\psi_{npc}(\widetilde{P}_n)\,,\,\psi_{npc}(P)) \to 0 \text{ (in probability), if } p/n \to 0,$$

(ii) $$d_2(\psi_{np}(\widetilde{P}_n)\,,\,\psi_{np}(P)) \to 0 \text{ (in probability),}$$

if $p^2/n \to 0$ *and* $d_2(\widehat{P}_n\,,\,P) = O_P(p^{-1/2})$.

Here $\widehat{P}_n$ *denotes the empirical distribution of the error variables, i.e.* $\widehat{P}_n(B) =$

$n^{-1} \sum_{i=1}^{n} I(\varepsilon_i \in B)$.

PROOF. We will show

$$E\, d_2^2(\widetilde{P}_n, \widehat{P}_n) \le \sigma^2(p+1)/n. \tag{1.1}$$

This implies (ii). For the proof of (i) note that $E\,\varepsilon_i^2 < +\infty$ implies $d_2(\widehat{P}_n, P) \to 0$ (in probability) (see Bickel and Freedman(1981)).

Proof of (1.1). As above assume $\mathbf{X}^T\mathbf{X} = I$ and define the residuals $\hat{\varepsilon}_i = Y_i - X_i^T \hat{\beta}$. We put $\hat{\varepsilon}_. = n^{-1} \sum_{j=1}^{n} \hat{\varepsilon}_j$ and $\varepsilon_. = n^{-1} \sum_{j=1}^{n} \varepsilon_j$. Then

$$\begin{aligned}
d_2^2(\widetilde{P}_n, \widehat{P}_n) &\le n^{-1} \sum_{i=1}^{n} ((\hat{\varepsilon}_i - \hat{\varepsilon}_.) - \varepsilon_i)^2 \\
&= n^{-1} \sum_{i=1}^{n} (\hat{\varepsilon}_i - \hat{\varepsilon}_. - \varepsilon_i + \varepsilon_.)^2 + \varepsilon_.^2 \\
&\le n^{-1} \sum_{i=1}^{n} (\hat{\varepsilon}_i - \varepsilon_i)^2 + \varepsilon_.^2 \\
&= n^{-1} \sum_{i=1}^{n} (\Sigma_{j=1}^{n} X_i^T X_j \varepsilon_j)^2 + \varepsilon_.^2 \\
&= n^{-1} \Sigma_{i,j,k=1}^{n} X_j^T X_i X_i^T X_k \varepsilon_j \varepsilon_k + \varepsilon_.^2 \\
&= n^{-1} \sum_{j,k=1}^{n} X_j^T X_k \varepsilon_j \varepsilon_k + \varepsilon_.^2 .
\end{aligned}$$

This implies:

$$E\, d_2^2(\widetilde{P}_n, \widehat{P}_n) \le n^{-1} \sigma^2 \Sigma_{j=1}^{n} \|X_j\|^2 + \sigma^2/n = \sigma^2(p+1)/n.$$

The corollary implies that bootstrap of linear contrasts $c^T(\hat{\beta} - \beta)$ works if $p/n = n^{-1} \Sigma_{i=1}^{n} \|X_i\|^2 \to 0$ whereas for asymptotic normality one has to assume $\max_{1 \le i \le n} \|X_i\|^2 \to 0$. This assumption for asymptotic normality means that the observation at every design point X_i has asymptotically negligible influence on the estimator $\hat{\beta}$, i.e. that there exists no leverage point

X_i. Note that bootstrap works also in the presence of leverage points as long as $p/n \to 0$.

We give two simple examples

EXAMPLE 1 (ONE WAY LAYOUT). Consider the model

$$Y_{ij} = \beta_i + \varepsilon_{ij} \qquad (i = 1, \dots, p; j = 1, \dots, T_i)$$

where the ε_{ij}'s are i.i.d. with mean 0. Then $H_{ii} = 1/T_i$. Therefore bootstrap works if $n/p \to +\infty$ whereas for asymptotic normality

$$\min_{1 \le i \le n} T_i \to +\infty$$

is necessary.

EXAMPLE 2 (POLYNOMIAL REGRESSION). Consider the model

$$Y_i = f(t_i) + \varepsilon_i \ (i = 1, \dots, n)$$

where f is a fixed function and $\varepsilon_1, \dots, \varepsilon_n$ are i.i.d. mean 0 random variables. The design points $t_i = t_{i,n}$ may depend on n. Consider a sequence of functions $f_1 = f_{1,n}, \dots, f_p = f_{p,n}$. For simplicity we assume that $f_1, \dots, f_p$ are orthonormal with respect to the empirical design measure, i.e.

$$< f_i , f_j >_n = \begin{cases} 0 & \text{for } i \neq j \\ 1 & \text{for } i = j, \end{cases}$$

where $< g, h>_n = n^{-1} \sum_{i=1}^n g(t_i) h(t_i)$.

The parameter $\beta = \beta_n$ is defined by

$$\sum_{i=1}^n (f(t_i) - \beta^T (f_1, \dots, f_p)^T (t_i))^2 = \min!$$

As above bootstrap works for $n/p \to +\infty$. For asymptotic normality of all linear contrasts of $\widehat{\beta}$ one has to assume the stronger condition

$$\max_{1 \le i \le n} n^{-1} \Sigma_{j=1}^{p} f_j^2(t_i) \to 0 .$$

2. Bootstrapping F - tests. Bootstrap may be used to estimate critical values of an F-test statistic. In this section we will show that bootstrap works in high - dimensional linear models under weak conditions. For linear subspaces $\mathbf{H}_0 \subset \mathbf{H}_1$ of $\mathbf{R}^p$ $(0 \in \mathbf{H}_0)$ we consider the test problem $\beta \in \mathbf{H}_0$ versus $\beta \in \mathbf{H}_1$. The F-test statistic for this test problem is defined by

$$T = \frac{\|\Pi_1 \mathbf{Y} - \Pi_0 \mathbf{Y}\|^2 / (p_1 - p_0)}{\|\mathbf{Y} - \Pi_1 \mathbf{Y}\|^2 / (n - p_1)}$$

where (for i=1,2) $\Pi_i \mathbf{Y}$ is the projection of $\mathbf{Y}$ onto $L_i = \{z \mid z_j = X_j^T b \text{ for a } b \in \mathbf{H}_i\}$ and p_i is the dimension of L_i. Denote the least squares estimator under the hypothesis $\beta \in \mathbf{H}_i$ by $\widehat{\beta}_i$. For the determination of critical values we want to estimate the distribution of T for parameters β in the hypothesis $\mathbf{H}_0$. Furthermore on the alternative $\beta \in \mathbf{H}_1 \backslash \mathbf{H}_0$ the bootstrap estimate of the distribution of T should converge to the distribution of T under a parameter β_0 which lies in the hypothesis $\mathbf{H}_0$. This would guarantee a satisfactory power of T when the bootstrap critical values are used.

As in Section 1 we consider the bootstrap procedure:

$$(\varepsilon_1^*, \dots, \varepsilon_n^*) \text{ is a resample from } \{(\hat{\varepsilon}_1 - \hat{\varepsilon}_.), \dots, (\hat{\varepsilon}_n - \hat{\varepsilon}_.)\} \tag{2.1}$$

where $\hat{\varepsilon}_i = Y_i - X_i^T \widehat{\beta}_1$. The bootstrap observations are defined as

$$\mathbf{Y}^* = \mathbf{X}\widehat{\beta}_0 + \varepsilon^*. \tag{2.2}$$

Then the bootstrap test statistic is defined as

$$T^* = \frac{\|(\Pi_1-\Pi_0)(\mathbf{Y}^*)\|^2 / (p_1-p_0)}{\|(I-\Pi_1)(\mathbf{Y}^*)\|^2 / (n-p_1)} .$$

We use the following Mallows distances

$$d_4(\mu,\nu) = \inf \{ (E\|X - Y\|^4)^{1/4} : \mathcal{L}(X) = \mu, \mathcal{L}(Y) = \nu \}$$

and

$$d(\mu,\nu) = \inf \{ (E[\inf(\|X-Y\|^2, 1)])^{1/2} : \mathcal{L}(X) = \mu, \mathcal{L}(Y) = \nu \}.$$

In the next theorem consistency of bootstrap of T and of $\|(\Pi_1-\Pi_0)(\mathbf{Y})\|^2$ is shown.

THEOREM 3. *Assume* $p_1/n \to 0$ *and*

(2.3) $\quad Y_i = X_i^T\beta + \varepsilon_i$

where $\beta \in \mathbf{H}_1$ *and* ε_i *are i.i.d. with* $E\,\varepsilon_i = 0$ *and* $E\,\varepsilon_i^4 < +\infty$. *Then*

(2.4) $\quad d(\mathcal{L}^*((p_1-p_0)^{1/2}\, T^*), \mathcal{L}((p_1-p_0)^{1/2}\, T^0)) \to 0$ *(in probability) where*

$$T^0 = \frac{\|(\Pi_1-\Pi_0)(\mathbf{Y}-\mathbf{X}\beta)\|^2 / (p_1-p_0)}{\|(I-\Pi_1)(\mathbf{Y})\|^2 / (n-p_1)},$$

(2.5) $\quad d(\mathcal{L}^*([\, n / (n-p_1) \,]\, \|(\Pi_1-\Pi_0)(\mathbf{Y}^*)\|^2 / (p_1-p_0)^{1/2}), \mathcal{L}(\|(\Pi_1-\Pi_0)(\mathbf{Y})\|^2 / (p_1-p_0)^{1/2}) \to 0$

(in probability).

Statement (2.5) makes sense because $(p_1-p_0)^{1/2}$ is the right norming factor : $\|(I-\Pi_1)(\mathbf{Y})\|^2 / (n-p_1) = E\,\varepsilon_i^2 + O_P(n^{-1/2})$ and $\|(\Pi_1-\Pi_0)(\mathbf{Y}-\mathbf{X}\beta)\|^2 / (p_1-p_0) = E\,\varepsilon_i^2 + O_P((p_1-p_0)^{-1/2})$ (see the proof of Theorem 3). If we omit the scaling factor $[\, n / (n-p_1) \,]$ in (2.5) then for the consistency of bootstrap in (2.5) we need to assume $\sqrt{p_1 - p_0}\, p_1/n \to 0$. The use of this scaling factor is equivalent to resampling from $\{[\, n / (n-p_1) \,]^{1/2} (\hat{\varepsilon}_1-\hat{\varepsilon}_.\,), \ldots,$

$[\, n / (n-p_1) \,]^{1/2} (\hat{\varepsilon}_n - \hat{\varepsilon}_.\,)\}$.

Note that if $\beta \in \mathbf{H}_0$ then $T = T^0$. Therefore bootstrap estimates the distribution of T consistently on the hypothesis $\mathbf{H}_0$. On the alternative $\mathbf{H}_1$ the bootstrap estimate converges to the distribution of T under a distribution with $\beta = 0 \in \mathbf{H}_0$.

This result may be compared with the necessary conditions for the validity of the approximation of $\mathcal{L}(T)$ by an F-distribution. It can be seen that for bounded p_1-p_0 it is necessary and sufficient that the maximal diagonal element of $(\Pi_1 - \Pi_0)$ converges to 0. Note that this condition is not necessary for the consistency of bootstrap.

3. Proof of Theorem 3. Suppose $\sum_{i=1}^n X_i X_i^T = I$ (w.L.o.G.) and define $X_{i,1}$ and $X_{i,2}$ as the projections of X_i onto $\mathbf{H}_1$ or onto $\mathbf{H}_0^{\perp} \cap \mathbf{H}_1$ (respectively). Then $\widehat{\beta}_1 = \sum_i X_{i,1} Y_i$ and $\sum_i \|X_{i,1}\|^2 = \text{spur} \sum_i X_{i,1} X_{i,1}^T = p_1$. Furthermore

$$(3.1) \quad \frac{1}{n-p_1} \|(I-\Pi_1)(\mathbf{Y})\|^2 = \frac{1}{n-p_1} \|\varepsilon\|^2 - \frac{1}{n-p_1} \|\Pi_1(\varepsilon)\|^2$$

$$= \frac{1}{n-p_1} \sum_{i=1}^n \varepsilon_i^2 - \frac{1}{n-p_1} \sum_{i,j=1}^n \varepsilon_i X_{i,1}^T X_{j,1} \varepsilon_j$$

$$= E\varepsilon_i^2 + O_P(1/\sqrt{n})$$

because of

$$\text{var}\Big(\frac{1}{n-p_1} \sum_{i \neq j} \varepsilon_i (X_{i,1}^T X_{j,1}) \varepsilon_j\Big) \leq \Big(\frac{1}{n-p_1}\Big)^2 2 \sum_{i \neq j} E\varepsilon_i^2 (X_{i,1}^T X_{j,1})^2 \varepsilon_j^2$$

$$\leq (n-p_1)^{-2} (E\varepsilon_i^2)^2 \, 2 \sum_i \|X_{i,1}\|^2$$

$$= O(p_1 (n-p_1)^{-2})$$

$$= o(n^{-1})$$

and

$$\mathrm{var}\Big(\frac{1}{n-p_1} \sum_i (1 - \|X_{i,1}\|^2) \varepsilon_i^2 = O((n-p_1)^{-2} \sum_i (1 - \|X_{i,1}\|^2)^2) = O(n^{-1}).$$

(Note that $\sum_i \|X_{i,1}\|^4 \leq \sum_i \|X_{i,1}\|^2 = p_1$ because of $\|X_{i,1}\| \leq 1$). Similarly one gets

$$(3.2)\quad \frac{1}{n-p_1} \|(I-\Pi_1)(\mathbf{Y}^*)\|^2 = \frac{1}{n-p_1} \|\varepsilon^*\|^2 - \frac{1}{n-p_1} \|\Pi_1(\varepsilon^*)\|^2$$

$$= E^*(\varepsilon_i^*)^2 + O_p(1/\sqrt{n})$$

$$= \frac{1}{n} \sum_{i=1}^n (\hat{\varepsilon}_i - \hat{\varepsilon}_.)^2 + O_p(1/\sqrt{n})$$

$$= n^{-1} \sum_{i=1}^n \hat{\varepsilon}_i^2 + O_p(1/\sqrt{n})$$

$$= n^{-1} \|(I-\Pi_1)(\varepsilon)\|^2 + O_p(1/\sqrt{n})$$

$$= (n - p_1) / n\, E\varepsilon_i^2 + O_P(1/\sqrt{n}),$$

where in the fourth equation the following has been used

$$(3.3)\quad E(\hat{\varepsilon}_.^2) = E((n^{-1}\sum_j \varepsilon_j - n^{-1}\sum_{i,j} (X_{i,1}^T X_{j,1})\varepsilon_j)^2)$$

$$= E(\varepsilon_i^2) n^{-2} \sum_j (1 - \sum_i (X_{i,1}^T X_{j,1}))^2$$

$$\leq 2\, E(\varepsilon_i^2) \left[n^{-2} \sum_{i,j,k} (X_{i,1}^T X_{j,1})(X_{j,1}^T X_{k,1}) + n^{-1} \right]$$

$$= 2\, E(\varepsilon_i^2) \left[n^{-2} \sum_{i,k} (X_{i,1}^T X_{k,1}) + n^{-1} \right]$$

$$\leq 2\, E(\varepsilon_i^2) \left[n^{-2} \Big(\sum_{i,k} (X_{i,1}^T X_{k,1})^2 n^2\Big)^{1/2} + n^{-1} \right]$$

$$= 2\, E(\varepsilon_i^2) \left[n^{-1} \Big(\sum_i \|X_{i,1}\|^2\Big)^{1/2} + n^{-1} \right]$$

$$= 2\, E(\varepsilon_i^2) (\sqrt{p_1} + 1) / n = o(1/\sqrt{n}).$$

We show now that

$$(3.4) \qquad d_4(\widehat{P}_n, \mathcal{L}^*(\varepsilon_i^*)) \to 0 \text{ (in probability).}$$

For (3.4) it suffices to show $n^{-1} \sum_j \left(\hat{\varepsilon}_. + X_j^T(\widehat{\beta}_1 - \beta)\right)^4 \to 0$. Because of (3.3) this follows from

$$(3.5) \qquad \frac{1}{n} \sum_j \left(X_j^T(\widehat{\beta}_1 - \beta)\right)^4 \to 0 \quad \text{(in probability).}$$

Proof of (3.5). $E\, n^{-1} \sum_j \left(X_j^T(\widehat{\beta}_1 - \beta)\right)^4 = E\, n^{-1} \sum_j \left(\sum_i X_{j,1}^T X_{i,1} \varepsilon_i\right)^4$

$$\le \text{const } n^{-1} \sum_{i,j,k} (X_{j,1}^T X_{i,1})^2 (X_{j,1}^T X_{k,1})^2$$

$$= \text{const } n^{-1} \sum_j \| X_{j,1} \|^4$$

$$\le \text{const } n^{-1} \sum_j \| X_{j,1} \|^2 \to 0.$$

Now because of (3.3) there exist (conditionally) independent η_i, η_i^* with $\mathcal{L}^*(\eta_i) = \mathcal{L}(\varepsilon_i)$, $\mathcal{L}^*(\eta_i^*) = \mathcal{L}^*(\varepsilon_i^*)$ and $E^*(\eta_i^2 - (\eta_i^*)^2)^2 \to 0$ (in probability) (Here E^* denotes the conditional expectation $E^*(\ldots) = E(\ldots \mid \mathbf{X},\mathbf{Y})$). Define $\mathbf{Z} = \mathbf{X}\beta + \eta$ and $Z^* = \mathbf{X}\beta + \eta^*$.

Then with $c_n = n / (n - p_1)$ one gets that $E\varepsilon_i^2 - c_n E^*(\varepsilon_i^*)^2 = O_p(1/\sqrt{n})$ (see (3.2)) and that

$$(3.6) \qquad E^*\Big[\frac{1}{\sqrt{p_1 - p_0}} \|(\Pi_1-\Pi_0)(\mathbf{Z} - \mathbf{X}\beta)\|^2 - \frac{1}{\sqrt{p_1 - p_0}} c_n \|(\Pi_1-\Pi_0)(\mathbf{Z}^*)\|^2\Big]^2$$

$$= E^* \frac{1}{p_1 - p_0} \Big\{ \sum_i \|X_{i,2}\|^2 (\eta_i^2 - c_n (\eta_i^*)^2)$$

$$+ \sum_{i \neq j} (X_{i,2}^T X_{j,2}) (\eta_i \eta_j - c_n \eta_i^* \eta_j^*) \Big\}^2$$

$$\le 2\,(p_1-p_0)^{-1}\,\{E^*[\textstyle\sum_i \ldots]^2 + E^*[\sum_{i\neq j} \ldots]^2\}$$

$$\le T_1 + T_2 + T_3$$

where

$$T_1 = \frac{2}{p_1-p_0}\sum_i E^*[\,(\ldots)^2\,] = \frac{2}{p_1-p_0}\sum_i E^*\|X_{i,2}\|^4\,(\eta_i^2 - c_n\,(\eta_i^*)^2)^2$$

$$\le 2E^*((\eta_i^2 - c_n\,(\eta_i^*)^2)^2$$

$$= o_P(1),$$

$$T_2 = \frac{2}{p_1-p_0}\,\{E^*[\textstyle\sum_i \ldots]\}^2 \le \frac{2}{p_1-p_0}\,(p_1-p_0)^2\{E^*\eta_i^2 - c_n\,E^*(\eta_i^*)^2\}^2$$

$$= O_P(\frac{p_1-p_0}{n})$$

$$= o_P(1),$$

$$T_3 = \frac{2}{p_1-p_0}E[\textstyle\sum_{i\neq j} \ldots]^2 \;\le 4\,(p_1 - p_0\,)^{-1}\sum_{i,j}\,(\,X_{j,2}^T X_{i,2})^2\,E^*(\eta_1\eta_2 - \eta_1^*\eta_2^*)^2$$

$$= 4\,E^*(\eta_1\eta_2 - \eta_1^*\eta_2^*)^2$$

$$= o_P(1).$$

Now (3.1), (3.2) and (3.6) imply the statement of the theorem.

Chapter 7. Bootstrapping robust regression.

1. Introduction. In the last chapter we have made extensive use of the simple (linear) structure of the model and of the estimate. As an example of a more complicated estimator we study in this chapter bootstrap of M-estimates $\hat{\beta}$ in linear models. As in the last chapter for each n we consider the linear model

$$Y_i = X_i^T\beta + \varepsilon_i \qquad (i = 1, \dots, n) \tag{1.1}$$

where the X_i's and β are p-dimensional vectors, the Y_i's are the observations and the ε_i's are i.i.d. errors distributed according to a distribution P. The X_i's and p may depend on n (see the discussion of this point in the last chapter). The content of this chapter is also contained in Mammen(1989a).

An M-estimator $\hat{\beta}$ or $(\tilde{\beta}, \hat{\sigma})$ is defined by

$$\sum_{i=1}^n X_i\, \psi\left(Y_i - X_i^T\hat{\beta}\right) = 0 \tag{1.2}$$

or by

$$\sum_{i=1}^n X_i\, \psi\left(\left(Y_i - X_i^T\tilde{\beta}\right)/\hat{\sigma}\right) = 0 \tag{1.3}$$

$$\sum_{i=1}^n \chi\left(\left(Y_i - X_i^T\tilde{\beta}\right)/\hat{\sigma}\right) = 0 \tag{1.4}$$

where $\psi, \chi : \mathbf{R} \to \mathbf{R}$ are given functions. For a discussion of robustness properties of M-estimates in linear models we refer to Hampel et al. (1986). But it should be remarked that several proposed estimates with high breakdown point are M-estimates with redescending ψ function where the solution of (1.2) or (1.3), (1.4) is chosen properly (see Rousseeuw and Yohai (1984), Yohai (1987), and Davies(1990)).

An overview of papers connected with the asymptotics of M-estimators in linear models with increasing dimension is contained in Portnoy (1984). The most general result is stated in Portnoy (1985). There it is assumed that the dimension p grows with n in such a way that

(1.5) $$p^{3/2} (\log n)^{3/2} / n \to 0 \quad \text{as} \quad n \to \infty .$$

Then, under some technical conditions on ψ and complicated assumptions about the design, it is shown that certain linear contrasts are approximately Gaussian. The conditions on the design are not fulfilled by ANOVA designs. For instance if the X_i are i.i.d. random variables the conditions hold in probability under strong assumptions on the tails of the distribution of the X_i.

The aim of this chapter is to show that bootstrap works under weaker conditions than the classical normal approximation. For this purpose we will give a stochastic expansion of the M-estimate $\widehat{\beta}$ (and $(\widetilde{\beta}, \widehat{\sigma})$) which holds under the condition that

(1.6) $$\max_{1 \le i \le n} H_{ii} \; n^{1/3} \; (\log n)^{2/3} \to 0$$

where H_{ii} is the i-th diagonal element of the hat matrix (see the last chapter). For the stochastic expansion of the M-estimate no further assumptions on the design are needed. The conditions used are weaker than in Portnoy (1985). There the conditions imply in particular that the design is balanced in a complicated sense, contains no leverage point, and that $\max_{1 \le i \le n} H_{ii} = O(p/n)$. Then (1.6) follows from (1.5).

Consistency of bootstrap of M-estimates in linear models has been shown by Shorack (1982) under the condition $\max_{1 \le i \le n} H_{ii} \, p^2 \to 0$. Under this condition the classical linear stochastic expansion of $\widehat{\beta}$ holds, i.e. $\| \widehat{\beta} - \beta - [\; E\psi' (\varepsilon_i)]^{-1} \sum_{i=1}^{n} X_i \, \psi(\varepsilon_i) \| = o_p(1)$ if the parametrisation β is chosen such that $\sum_{i=1}^{n} X_i \, X_i^T = I$. This implies also the validity of the normal approximation (see also Huber, 1981). Under our assumption (1.6) the linear (and also the higher - order) stochastic expansion of fixed dimension p does not hold in general. Furthermore the classical normal approximation is not valid. This shows also that an approach based on higher - order Edgeworth expansions is different from our approach of dimension asymptotics. An analysis based on Edgeworth expansions seems to be of a limited value in the case of a high - dimensional model. This is connected with the empirical fact that Edgeworth expansions usually produce significant improvements only when the Gaussian approximation is not too inaccurate. And clearly the assumptions necessary for Edgeworth expansions are stronger than those for Gaussian approximations. Thus it seems more promising to weaken the assumptions of the classical asymptotic approach in such a way that more features of the finite sample become asymptotically relevant.

Another example for dimension asymptotics is provided by sparse contingency tables. Standard asymptotics do not take into account the sparsity of the table and may therefore lead to inaccurate approximations. Koehler (1986) proposes another approach to the asymptotics of goodness - of - fit statistics testing the overall fit of a log-linear model. This approach is based on an increasing number of categories and leads to approximations which are more accurate for sparse tables, as can be seen in simulation studies. Ehm (1986, 1991), generalising results of Haberman (1977a,b), has used a dimension-asymptotics approach to study the behaviour of ML-estimates of loglinear models in sparse and unbalanced tables. In particular he has derived diagnostics for the accuracy of the usual Gaussian approximation. For loglinear models with closed form ML-estimates Sauermann (1986, 1989) shows the superiority of the parametric bootstrap over Gaussian approximations in the case of sparse tables. This is done using dimension asymptotics combined with extensive Monte Carlo simulations. For arbitrary exponential families with increasing number of parameters the asymptotic behaviour of the ML-estimate and of the likelihood ratio test has been studied by Portnoy (1988). For an in spirit related asymptotic treatment of bootstrap for nonparametric change point problems see Dümbgen (1991a).

2. Bootstrapping M - estimates. For simplicity we assume

$$\sum_{i=1}^{n} X_i X_i^T = I. \tag{2.1}$$

Note that this assumption corresponds to $\sqrt{n}$ renorming of the parameter in a location model (see Chapter 6). We describe first the bootstrap algorithm for $\hat{\beta}$. Assume

$$E\ \psi(\varepsilon_i) = 0. \tag{2.2}$$

To estimate the distribution of $\hat{\beta}$ the bootstrap proceeds as follows. Define the residuals $\hat{\varepsilon}_i = Y_i - X_i^T \hat{\beta}$ $(i = 1, \dots, n)$ and let $\tilde{P}_n$ denote the empirical distribution of $\hat{\varepsilon}_1, \dots, \hat{\varepsilon}_n$. Now generate n i.i.d. observations $\varepsilon_1^*, \dots, \varepsilon_n^*$ distributed according to $\tilde{P}_n$. Define $Y_i^* = X_i^T \hat{\beta} + \varepsilon_i^*$ $(i = 1, \dots, n)$ and $\hat{\beta}^*$ as the solution of the following M-equation

$$\sum_{i=1}^{n} X_i\ \tilde{\psi}\left(Y_i^* - X_i^T \hat{\beta}^* \right) = 0, \tag{2.3}$$

where $\widetilde{\psi}(u) = \psi(u) - \int \psi \, d\widetilde{P}_n$. Note that the ψ function is changed such that the bootstrap analogon of (2.2) holds, i.e. $E^* \widetilde{\psi}(\varepsilon_i^*) = 0$. In the case of $(\widetilde{\beta}, \widehat{\sigma})$ we assume for $\sigma > 0$

$$\text{(2.4)} \qquad E\,\psi(\varepsilon_i / \sigma) = 0,$$

$$\text{(2.5)} \qquad E\,\chi(\varepsilon_i / \sigma) = 0\,.$$

For simplicity suppose $\sigma = 1$. Now n i.i.d. observations $\varepsilon_1^*, \dots, \varepsilon_n^*$ are generated which are distributed according to the empirical distribution of the residuals $Y_i - X_i^T \widetilde{\beta}$ (i = 1, ... , n) . Define $Y_i^* = X_i^T \widetilde{\beta} + \varepsilon_i^*$ (i = 1, ... , n) and $(\widetilde{\beta}^*, \widehat{\sigma}^*)$ as the solution of the following M-equations

$$\text{(2.6)} \qquad \sum_{i=1}^n X_i \, \overline{\psi}\left(\left(Y_i^* - X_i^T \widetilde{\beta}^* \right) / \widehat{\sigma}^* \right) = 0,$$

$$\text{(2.7)} \qquad \sum_{i=1}^n \overline{\chi}\left(\left(Y_i^* - X_i^T \widetilde{\beta}^* \right) / \widehat{\sigma}^* \right) = 0,$$

where $\overline{\psi}(u) = \psi(u) - \int \psi(v / \widehat{\sigma}_n) \, \widetilde{P}_n(dv)$ and $\overline{\chi}(u) = \chi(u) - \int \chi(v / \widehat{\sigma}_n) \, \widetilde{P}_n(dv)$. The solutions of the M-equations (1.2) (0 [(1.3), (1.4)], (2.3), or [(2.6), (2.7)] respectively) may not be unique. For instance for redescending ψ the number of solutions of (1.2) may tend to infinity. Some estimates with high breakdown point which have been discussed in the last years (see Rousseeuw, Yohai (1984) and Yohai (1987)) are M-estimates with redescending ψ function where the solution of the M-equation is cleverly chosen. In the next section we will see that the solution of the M-equations (1.2) (of (1.3), (1.4) respectively) is unique on $\{ b: \max_{1 \le i \le n} |X_i^T (b - \beta)| \le \delta \}$ (on $\{(b,s): \max_{1 \le i \le n} |X_i^T (b - \beta)| \le \delta, \ |s - \delta| \le \delta \}$ respectively) for δ small enough. For simplicity we define here $\widehat{\beta}$ (and $\widetilde{\beta}$, $\widehat{\sigma}$)) as the solution in this set. Clearly this choice is only theoretically possible because β, σ, and δ are unknown. For the application of our results to estimates (which are based on a really possible choice of the solution) one has to show additionally the consistency of these estimates, i.e.

$$\max_{1 \le i \le n} |X_i^T (\widehat{\beta} - \beta))| \to 0 \quad \text{(in probability) or}$$

$$\max_{1 \le i \le n} |X_i^T (\widetilde{\beta} - \beta)| \to 0 \quad \text{and } |\widehat{\sigma} - \sigma| \to 0 \ \text{(in probability).}$$

In the bootstrap world we define $\widehat{\beta}^*$ and $(\widetilde{\beta}^*, \widehat{\sigma}^*)$ as the solution with smallest distance to $\widehat{\beta}$ (or to $(\widetilde{\beta}, \widehat{\sigma})$ respectively) (where the used distance may be for instance $\Delta(\beta_1, \beta_2) = \max_{1 \le i \le n} |X_i^T(\beta_1 - \beta_2)|$ or $\Delta(\beta_1, \beta_2) \vee |\sigma_1 - \sigma_2|$ respectively). This choice of $\widehat{\beta}^*$ and $(\widetilde{\beta}^*, \widehat{\sigma}^*)$ can be calculated very fast because one has to search for the solutions of the M-equations only in a *local* neighborhood of $\widehat{\beta}$ or $(\widetilde{\beta}, \widehat{\sigma})$. This is in contrast to the calculation in the real world where the calculation of high breakdown point estimators uses highly computer intensive *global* searches. This difference means that bootstrap of even these computer intensive methods can be calculated in a reasonable amount of computer time.

We will show consistency of bootstrap under the following smoothness conditions on ψ and χ.

(2.8) *ψ is a bounded function with three bounded derivatives fulfilling* $E\,\psi(\varepsilon_i) = 0$ *and* $d = E\,\psi'(\varepsilon_i) > 0$.

(2.9) *χ is a bounded function with three bounded derivatives fulfilling* $E\chi(\varepsilon_i) = 0$. *Furthermore* $x^3 \chi^{(3)}(x)$ *and* $x^3 \psi^{(3)}(x)$ *are bounded and* $E\,\psi'(\varepsilon_i)\,\varepsilon_i\, E\chi'(\varepsilon_i) < E\,\psi'(\varepsilon_i)\, E\chi'(\varepsilon_i)\,\varepsilon_i$.

THEOREM 1. *Suppose* (2.8) *and*

(2.10) $\max_{1 \le i \le n} H_{ii}\, n^{1/3} (\log n)^{2/3} \to 0.$

Furthermore assume w.l.o.g. (2.1) . *Then for* $c_n \in \mathbf{R}^p$ *with* $\| c_n \|$ *bounded*

(2.11) $$d_M(\mathcal{L}(c_n^T(\widehat{\beta} - \beta)), \mathcal{L}^*(c_n^T(\widehat{\beta}^* - \widehat{\beta}))) \to 0 \text{ (in probability),}$$

(2.12) $$d_M(\mathcal{L}(c_n^T(\widetilde{\beta} - \beta), (\widehat{\sigma} - 1), \mathcal{L}^*(c_n^T(\widetilde{\beta}^* - \widetilde{\beta}), \widehat{\sigma}^* - \widehat{\sigma})) \to 0$$

(in probability), if (2.9) *holds additionally.*

Assume additionally

(2.13) $$p^2/n \to 0$$

(2.14) $$\psi'(x) = 0 \quad \textit{for} \quad |x| \textit{ large enough}$$
$$\textit{or } d_2(P, \widehat{P}_n) = o_p(1/\sqrt{p}).$$

Then

(2.15) $$d_M(\mathcal{L}(\widehat{\beta} - \beta), \mathcal{L}^*(\widehat{\beta}^* - \widehat{\beta})) \to 0 \textit{ (in probability)},$$

(2.16) $$d_M(\mathcal{L}(\widetilde{\beta} - \beta, \widehat{\sigma} - 1), \mathcal{L}^*(\widetilde{\beta}^* - \widetilde{\beta}, \widehat{\sigma}^* - \widehat{\sigma})) \to 0$$

(in probability), if (2.9) *holds additionally.*

Here for two probability measures μ *and* ν *the quantity* $d_M(\mu, \nu)$ *denotes the following modification of the Mallows distance*

$$d_M(\mu,\nu) = \inf \{ [E(\| U - V \|^2 \wedge 1)]^{1/2} : \mathcal{L}(U) = \mu, \ \mathcal{L}(V) = \nu \}.$$

$d_2(\mu, \nu)$ *denotes the Mallows distance (see Chapter 6).*

In particular, Theorem 1 indicates that bootstrap works in cases where the usual asymptotic approach fails. The bootstrap estimates the bias of a linear contrast consistently - also in cases where the bias tends to ∞ (see Mammen, 1989a). If design points are leverage points it can happen that $p^2/n \to 0$ but that $\max_{1 \le i \le n} H_{ii} \cdot p$ does not converge to 0. Then the distribution of $\widehat{\beta}$ cannot be approximated by a Gaussian distribution (measured by the Mallows distance d_M). Nevertheless the bootstrap estimate of the distribution of $\widehat{\beta}$ is consistent. The proof of Theorem 1 is based on the fact that analogous stochastic expansions hold for $\widehat{\beta}$ and $\widehat{\beta}^*$. A similar approach has been used by Sauermann (1986, 1989) in the case of large sparse contingency tables. Also there the bootstrap estimates not only the distribution of the first order linear term of the stochastic expansion of an estimator but also the distribution of the higher - order terms. This phenomenon has also been recognized in studies of the bootstrap which are based on higher - order Edgeworth expansions (See for instance Beran (1984a, 1987, 1988), Hall(1986a,1988), and Chapter 5.).

In Figures 1a, ..., 2c the results of a small simulation study are displayed. We have estimated the real distribution of two linear contrasts and of a χ^2 test statistic by 1000 Monte Carlo simulations. Furthermore one bootstrap estimate based on the first observation sample is shown. In the bootstrap resampling 1000 replications are used. In the figures kernel estimates of the real density and kernel density estimates based on the bootstrap estimate of the distribution are compared with the asymptotic normal density (the asymptotic χ^2 density, respectively). The sample size n is 100 and the dimension p is 11. The distribution of the error variables ε_i is 1/2 N(1/2, $(1.2)^2$) + 1/2 N(–1/2, $(0.7)^2$) (as in the simulations of Chapter 1 (Model C) and in Chapter 5). The density of the error distribution is plottted in Figure 3. We have used the following Huber ψ function

$$\psi(x) = (x \wedge 1) \vee (-1).$$

The design vector X_i are generated from independent N(0,1) variables $U_{i,j}$ using the following shifts

MODEL A:

$$X_{i,j} = \begin{cases} 1 & \text{for } j=1 \text{ and } i=1, \dots, 100, \\ U_{i,j}+8 & \text{for } j=2 \text{ and } i=1, \dots, 20, \\ U_{i,j}-5 & \text{for } j=3 \text{ and } i=11, \dots, 30, \\ U_{i,j} & \text{else.} \end{cases}$$

MODEL B:

$$X_{i,j} = \begin{matrix} 1 & \text{for } j = 1 \text{ and } i = 1,\dots,100, \\ U_{i,j} + 8 & \text{for } j = 2 \text{ and } i = 1,\dots, 20, \\ U_{i,j} - 5 & \text{for } j = 3 \text{ and } i = 11,\dots,30, \\ U_{i,j} + 6 & \text{for } j = 4 \text{ and } i = 31,\dots,50, \\ U_{i,j} & \text{else.} \end{matrix}$$

Note that the design vectors lie in model A (model B) in five (six, respectively) strongly separated clusters. The following two linear contrasts $c^T(\hat{\beta} - \beta)$ have been simulated: $c = c_1 = (0, 1, 0, \dots, 0)^T$ and $c = c_2 = (0, \gamma_2, \dots, \gamma_{11})^T$ where $\gamma_j = \sum_{i=1}^{100} X_{i,j}$. Furthermore the χ^2 test of the following null hypothesis has been simulated: There exists a β_1 such that E $\psi(Y_i - \beta_1) = 0$, i.e. $\beta_2 = \dots = \beta_{11} = 0$. Up to a factor this test statistic has an asymptotic χ^2 distribution with 10 degrees of freedom. The figures show that bootstrap produces significantly better approximations than the normal approximation. Bootstrap produces accurate estimates of the variance of the linear contrasts and of the χ^2 test statistic whereas the asymptotic variance is too small in all cases. Furthermore bootstrap estimates the local wiggles of the underlying densities with surprising accuracy. This advantage of bootstrap cannot be explained with the help of Edgeworth expansions. An asymptotic analysis based on Edgeworth expansions would suggest that bootstrap is preferable because it corrects the normal approximation in respect of skewness and kurtosis. But this phenomenon is not visible here.

3. Stochastic expansions of the M-estimates. In this section we state a stochastic expansion for the M-estimate $\widehat{\beta}$. For an analogous result for $(\widetilde{\beta},\widehat{\sigma})$ we refer to Mammen (1989a).

THEOREM 2. *Assume* (2.1) , (2.8) , (2.10) . *Then for* δ *small enough there exists with probability tending to one a unique solution* $\widehat{\beta}$ *of* $\sum_{i=1}^n X_i\, \psi(Y_i - X_i^T \widehat{\beta}) = 0$ *in* $\{b : \max_{1\le i\le n} X_i^T (b-\beta) \le \delta\}$. *For* $\widehat{\beta}$ *the following expansion holds*

(3.1) $\|\widehat{\beta} - \widehat{\beta}_1\| = o_P(1)$, *where*

(3.2) $$\widehat{\beta}_1 = A^{-1} \widehat{\beta}_0 - A^{-1} B A^{-1} \widehat{\beta}_0 + 1/2\, A^{-1} \sum_{i=1}^n X_i (X_i^T A^{-1}(\widehat{\beta}_0 - E\, \widehat{\beta}_0))^2 \psi''(\varepsilon_i - X_i^T a) + a,$$

(3.3) $$\widehat{\beta}_0 = \sum_{i=1}^n X_i\, \psi(\varepsilon_i - X_i^T a),$$

(3.4) $$A = \sum_{i=1}^n X_i X_i^T\, E\, \psi'(\varepsilon_i - X_i^T a),$$

(3.5) $$B = \sum_{i=1}^n X_i X_i^T (\psi'(\varepsilon_i - X_i^T a) - E\, \psi'(\varepsilon_i - X_i^T a)).$$

Furthermore for a fixed constant C *the constant* a *(the asymptotic bias) is a solution of*

(3.6) $$\|a\| \le C\sqrt{\kappa p}$$

(3.7) $$E\widehat{\beta}_1 = a.$$

To prove Theorem 2 we will show that $\| \sum_{i=1}^n X_i\, \psi(\varepsilon_i - X_i^T \widehat{\beta}_1)\|$ is small. Then the Newton-Kantorowitsch Theorem implies that there exists a solution of (1.2) in the neighborhood of $\widehat{\beta}_1$. The main idea of the proof is to use only properties of $\widehat{\beta}_1$ and not of $\widehat{\beta}$. The proofs of Theorems 1 and 2 are given in the next section.

For p fixed, and $\kappa = \max_{1 \le i \le n} H_{ii} \to 0$ the M-estimate $\widehat{\beta}$ is asymptotically distributed according to $N(0, E\psi^2(\varepsilon_i) / E\psi'(\varepsilon_i))^2 \cdot I_p)$. Only in the case a = 0 does the stochastic expansion (3.1) coincide with the usual higher - order expansion for fixed dimension p. But under the assumptions of the theorem it may occur that $\| a \| \to \infty$. In (3.1) the norm of the first term is of order $\sqrt{p}$ and the norm of the second and the third term is of order $\sqrt{\kappa p}$. For balanced designs $\sqrt{\kappa p}$ is of order $p/\sqrt{n}$ because in this case $\kappa = \max_{1 \le i \le n} H_{ii}$ is of order p/n (under the norming condition (2.1)).

As in Huber (1981) and Portnoy (1985) the techniqual differentiability conditions assumed are not fulfilled by some commonly used ψ functions. Such smoothness conditions are necessary for a treatment of M-estimates which is based on a stochastic expansion. As indicated by Portnoy (1985) an alternative approach may be based on a study of the joint density of the M-estimate. Then the conditions on ψ could be relaxed by assuming smoothness of the density of the ε_i's.

An theorem analogous to Theorem 2 can be formulated for the case that the scale is estimated simultanously (see Mammen,1989a).

Theorem 2 can be used to prove asymptotic normality for linear contrasts.

THEOREM 3. *Assume* (2.1), (2.8), (2.10). *Then for* $c_n \in \mathbf{R}^p$ *with* $\| c_n \|$ *bounded*

$$[\, \| c_n \|^2 \, E\, \psi^2(\varepsilon_i) / (E\psi'(\varepsilon_i))^2 \,]^{-1/2} \, [\, c_n^T \widehat{\beta} - c_n^T (\beta + a) \,]$$

converges in distribution to N(0,1).

The bias $c_n^T a$ can be of order $\sqrt{\kappa p}$ (the order of $\| a \|$) which may tend to infinity under the assumptions of the theorem.

4. Proofs. We will only give sketches of the proofs of Theorems 1, 2, and 3. For a detailed proof see Mammen (1989a). Put $\kappa = \max_{1 \le i \le n} H_{ii} = \max_{1 \le i \le n} \| X_i \|^2$ (see (2.1)).

SKETCH OF THE PROOF OF THEOREM 2.

FIRST STEP. For C large enough there exists a solution a of (3.6), (3.7).

PROOF. Define g: $\mathbf{R}^p \to \mathbf{R}^p$ exactly as $E\,(A\,(\widehat{\beta}_1 - a) / E\psi'(\varepsilon_i) + a)$ (see (3.2) – (3.5)), but as a function of a. Then one can show, if C is chosen large enough, that

$$\| b \| \le C\sqrt{\kappa p} \text{ implies } \| g(b) \| \le c\sqrt{\kappa p}\,.$$

Then from Brouwer's fixed point theorem (Dunford and Schwartz (1958)) it follows that there exists an a with (3.6) and g(a) = a, i.e. (3.7).

SECOND STEP. In this step we show that $\| \Sigma X_i\, \psi(\varepsilon_i - X_i^T \widehat{\beta}_1 \|$ is small and we follow from this fact that $\widehat{\beta}_1$ must lie near to a solution of $\Sigma X_i\, \psi\,(\varepsilon_i - X_i^T \widehat{\beta}\,) = 0$. This can for instance be done by the application of a Newton-Kantorowitsch Theorem. We cite a version which will be used here (Kantorowitsch and Akilow, 1964).

THEOREM (NEWTON-KANTOROWITSCH). *Assume, that a function* G: $\mathbf{R}^p \to \mathbf{R}^p$ *has two continuous derivatives for* $\| x - x_0 \| \le r$. *Furthermore assume*

$$\Gamma := (G'(x_0))^{-1} \textit{ exists,} \tag{4.1}$$

$$\| \Gamma G(x_0) \| \le \eta, \tag{4.2}$$

$$\| \Gamma G''(x) \| \le \lambda \text{ for } \| x - x_0 \| \le r, \tag{4.3}$$

$$h = \lambda \cdot \eta \le 1/2, \tag{4.4}$$

$$r_0 := \frac{1 - \sqrt{1 - 2h}}{h}\,\eta \le r. \tag{4.5}$$

Then the equation G(x) = 0 *has a solution* x* *with*

$$\| x^* - x_0 \| \le r_0.$$

This theorem will be apllied as follows. We set $G(\theta) = \sum_{i=1}^n X_i \psi\left(\varepsilon_i - X_i^T\theta\right)$.

For the proof of Theorem 2 it suffices to show

$$\text{(4.6)} \qquad \| G'(\widehat{\beta}_1) - E\psi'(\varepsilon_i)\, I \| = o_p(1),$$

$$\text{(4.7)} \qquad \| G(\widehat{\beta}_1) \| = o_p(1),$$

$$\text{(4.8)} \qquad \sup_{\theta \in \mathbb{R}^p} \| G''(\theta) \| = o_p(1).$$

The essential point in the proof of (4.6) and (4.7) is the proof of

$$\text{(4.9)} \qquad \max_{1 \le i \le n} \left| X_i^T(\widehat{\beta}_1 - a) \right| = O_P\left(\sqrt{\kappa}\,\sqrt{\log n}\right).$$

Then (4.6) and (4.7) can be proved on the basis of Taylor expansions. For details and the proof of (4.9) see Mammen (1989a). Note that $X_i^T(\widehat{\beta}_1 - a)$ is a cubic form for every $i \le n$. The proof of (4.8) is simple:

$$\| G''(\theta) \| = \sup_{\substack{\|\tau_1\| \le 1 \\ \|\tau_2\| \le 1 \\ \|\tau_3\| \le 1}} \sum_{i=1}^n (X_i^T \tau_1)(X_i^T \tau_2)(X_i^T \tau_3)\, \psi''(\varepsilon_i - X_i^T \theta)$$

$$\le \text{const.} \sup_{\tau_1, \tau_2, \tau_3} \sqrt{\sum_{i=1}^n (X_i^T \tau_1)(X_i^T \tau_2) \sum_{i=1}^n (X_i^T \tau_3)^2}$$

$$\le \text{const.} \sup_{\|\tau_1\| \le 1} \sup_{1 \le i \le n} | X_i^T \tau_1 | = 0\,(\sqrt{\kappa}).$$

SKETCH OF THE PROOF OF THEOREM 3. Theorem 3 follows from

$$\text{(4.10)} \qquad c_n^T(\widehat{\beta}_1 - E\widehat{\beta}_1) - c_n^T(\widehat{\beta}_0 - E\widehat{\beta}_0) = o_P(1).$$

For the proof of (4.10) see Mammen (1989a).

SKETCH OF THE PROOF OF THEOREM 1. For simplicity of notation define a sample of n (conditionally) independent identically distributed random variables $R_1, \dots, R_n$ with conditional distribution $\mathcal{L}^*(R_i) = P$. Put $R_i^* = \widetilde{F}_n^{-1} F(R_i)$ where F is the distribution function of R_i (and ε_i) and $\widetilde{F}_n$ is the empirical distribution function of the residuals $Y_i - X_i^T \widehat{\beta}$ $(i = 1, \dots, n)$. Then $\mathcal{L}^*(R_i^*) = \widetilde{P}_n$.

Denote the empirical distribution of $\varepsilon_1, \ldots, \varepsilon_n$ by $\widehat{P}_n$. Then because of $\|\widehat{\beta} - \beta\|^2 = O_p(p)$ (see the proof of Theorem 2 in Mammen (1989a)) the Mallows distance $d_2(\widetilde{P}_n, \widehat{P}_n)$ can be bounded by

(4.11) $$d_2^2(\widetilde{P}_n, \widehat{P}_n) \le n^{-1} \sum_{i=1}^n \left(X_i^T(\widehat{\beta} - \beta)\right)^2$$

$$= n^{-1} \|\widehat{\beta} - \beta\|^2$$

$$= O_P(p/n).$$

Define $0 < \tau \le +\infty$ as $\tau = \inf\{t : \psi'(x) = 0 \text{ for } |x| \ge t\}$. Put $\delta^2 = E(R_i - R_i^*)^2 1(|R_i| \le 2\tau)$. Then under the assumption of (2.13) and (2.14)

(4.12) $$\delta = o_P(1/\sqrt{p}).$$

(4.12) follows from (4.11) and

(4.13) $$\int_0^1 \left(\widehat{F}_n^{-1}(x) - F^{-1}(x)\right)^2 \; 1(|F^{-1}(x)| \le D)\, dx$$

$$= O_P(1) \int_0^1 \left|\widehat{F}_n^{-1}(x) - F^{-1}(x)\right| \; 1(|F^{-1}(x)| \le D)\, dx$$

$$= O_P(1) \int_{-D+O_P(1)}^{D+O_P(1)} \left|\widehat{F}_n(x) - F(x)\right| dx$$

$$= O_P(n^{-1/2})$$

for every $D < +\infty$. Here $\widehat{F}_n$ denotes the empirical distribution function of $\varepsilon_1, \ldots, \varepsilon_n$.

For simplicity we abuse the notation slightly and identify $\varepsilon_i = R_i$ and $\varepsilon_i^* = R_i^*$. Furthermore we define $\widehat{\beta}$, $\widehat{\beta}_1$, $\widehat{\beta}_0$, A, B, and a as in Theorem 2. Then one can show, that

(4.14) $\|\widehat{\beta}^* - \widehat{\beta}_1^*\| = o_P(1)$

where

$$(4.15)\qquad \widehat{\beta}_1^{\,*} = A^{*-1}\,\widehat{\beta}_0^{\,*} - A^{*-1}\,B^*\,A^{*-1}\,\widehat{\beta}_0^{\,*}$$

$$+\frac{1}{2}A^{*-1}\sum_{i=1}^n X_i\left[X_i^T A^{*-1}\left(\widehat{\beta}_0^{\,*} - E^*\widehat{\beta}_0^{\,*}\right)\right]^2 \psi''(\varepsilon_i^* - X_i^T a^*) + a^*.$$

Furthermore

$$(4.16)\qquad \widehat{\beta}_0^{\,*} = \sum_{i=1}^n X_i\,\widetilde{\psi}(\varepsilon_i^* - X_i^T a^*),$$

$$(4.17)\qquad \widetilde{\psi}(x) = \psi(x) - \int \psi(u)\,\widetilde{P}_n(du),$$

$$(4.18)\qquad A^* = \sum_{i=1}^n X_i X_i^T E^*\psi'(\varepsilon_i^* - X_i^T a^*),$$

$$(4.19)\qquad B^* = \sum_{i=1}^n X_i X_i^T\,[\psi'(\varepsilon_i^* - X_i^T a^*) - E^*\psi'(\varepsilon_i^* - X_i^T a^*)]$$

and for a large enough fixed C the variable a^* is chosen such that

$$(4.20)\qquad \| a^* \| \le C\sqrt{\kappa p} \quad\text{and}\quad E^*\widehat{\beta}_1^{\,*} = a^*.$$

After these preliminaries we come to the proof of Theorem 1. We sketch only the proof of (2.11) and (2.15). The proof of (2.12) and (2.16) is similar.

Proof of (2.11). With the arguments given in the proof of Theorem 3 one sees that for (2.11) it suffices to show:

$$(4.21)\qquad \| a^* - a \| = O_P\left(\sqrt{\kappa p^2/n}\right).$$

Note that because of $\kappa = \max_{1\le i\le n} H_{ii} \ge p/n$ (This follows from (2.1) because of $\sum_{i=1}^n H_{ii} = \sum_{i=1}^n \|X_i\|^2 = p$) and $\kappa n^{1/3} \to 0$ (see (1.6))

$$(4.22)\qquad \kappa p^2/n \le \kappa^3 n \to 0.$$

To prove (4.21) one shows

(4.23) $$\| g(a^*) - g(a) \| = O_P\left(\sqrt{\kappa p^2 / n}\,\right),$$

(4.24) $$\| g'(b) - I \| = o(1)$$

uniformly in $\| b \| \leq \text{const} \sqrt{\kappa p}$ (g is defined at the beginning of the proof of Theorem 2.).

Proof of (2.15). First note that

$$d_2^2\left(\mathcal{L}(\hat{\beta}_0 - E\,\hat{\beta}_0),\ \mathcal{L}^*(\hat{\beta}_0^* - E^*\,\hat{\beta}_0^*)\right)$$

$$\leq \sum_{i=1}^n \| X_i \|^2\ E^*\left[\psi(R_i - X_i^T a) - E^*\,\psi(R_i - X_i^T a)\right.$$

$$\left. - \tilde{\psi}\left(R_i^* - X_i^T a^*\right) + E^*\,\tilde{\psi}\left(R_i^* - X_i^T a^*\right)\right]^2$$

$$\leq \text{const.} \sum_{i=1}^n \| X_i \|^2\ \left[\delta^2 + (X_i^T (a - a^*))^2\right]$$

$$\leq \text{const.}\left[p\,\delta^2 + \kappa \| a - a^* \|^2\right].$$

With similar arguments for the other terms of $\hat{\beta}_1$ and $\hat{\beta}_1^*$ one can show

(4.25) $$d_2^2\left(\mathcal{L}(\hat{\beta}_1 - E\,\hat{\beta}_1),\ \mathcal{L}^*(\hat{\beta}_1^* - E^*\,\hat{\beta}_1^*)\right)$$

$$\leq \text{const.}\left[p \cdot \delta^2 + \kappa \| a - a^* \|^2 + p \| A^{-1} - A^{*-1} \|^2 + \| a - a^* \|^2\right].$$

Now because of (4.12) the first term converges to zero in probability. The convergence of the second and the fourth term follows from (4.21) . For the proof of

$$d_2\left(\mathcal{L}(\hat{\beta}_1 - E\,\hat{\beta}_1),\ \mathcal{L}^*(\hat{\beta}_1^* - E^*\,\hat{\beta}_1^*)\right) = o_p(1)$$

it remains to bound the third term in (4.25)

$$p \| A^{-1} - A^{*-1} \|^2$$

$$\leq p \, \| A^{-1} \|^2 \, \| A^{*-1} \|^2 \, \| A - A^* \|^2$$

$$= O_P(p) \, [\sup_{\|\theta\| \leq 1} \sum_{i=1}^n (X_i^T \theta)^2 \, [\, E^* \, | R_i - R_i^* | \, 1 \, (| R_i | \leq 2 \tau)$$

$$+ | X_i^T (a - a^*) |]^2 \,]$$

$$= O_P(p) \, [\, \delta^2 + \kappa \, \| a - a^* \|^2 \,]$$

$$= o_P(1) \; + \; O_p(\kappa^2 \, p^3 / n)$$

$$= \; o_P(1)$$

where (4.12) and (4.21) have been used. Note that $\kappa^2 \, p^3 / n = o(\kappa^2 \, p) = o(\kappa^3 \, n) = o(1)$ because of (2.13), $\kappa = \max_{1 \leq i \leq n} H_{ii} \geq p/n$, and (4.22).

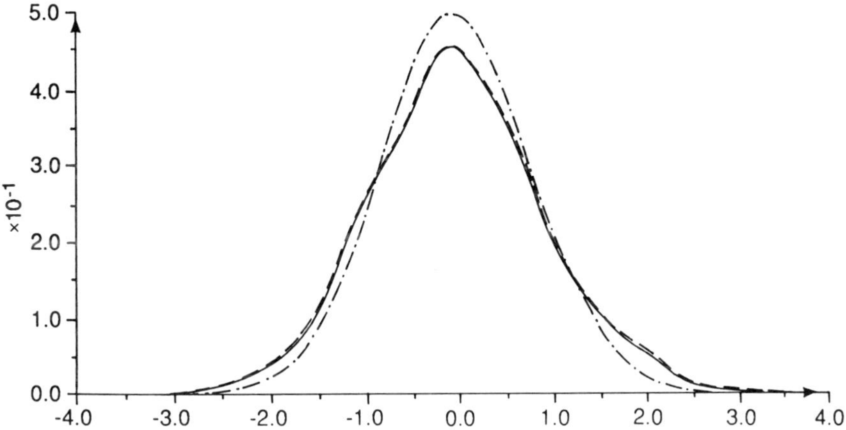

Figure 1a. *Kernel density estimates for the linear contrast* $c_1^T(\hat{\beta} - \beta)$ *based on 1000 pseudo samples* (———) *and based on one bootstrap estimate with 1000 replications* (— — —) *and the asymptotic normal density* (— · — · —) *(Model A)*

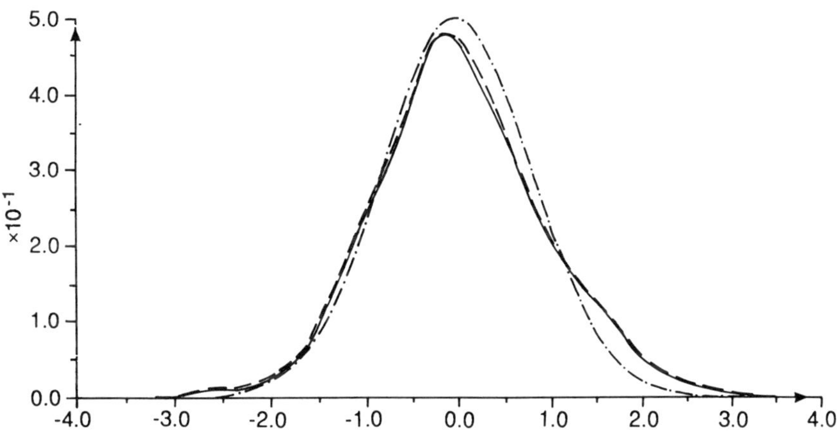

Figure 1b. *Kernel density estimates for the linear contrast* $c_2^T(\hat{\beta} - \beta)$ *based on 1000 pseudo samples* (———) *and based on one bootstrap estimate with 1000 replications* (— — —) *and the asymptotic normal density* (— · — · —) *(Model A)*

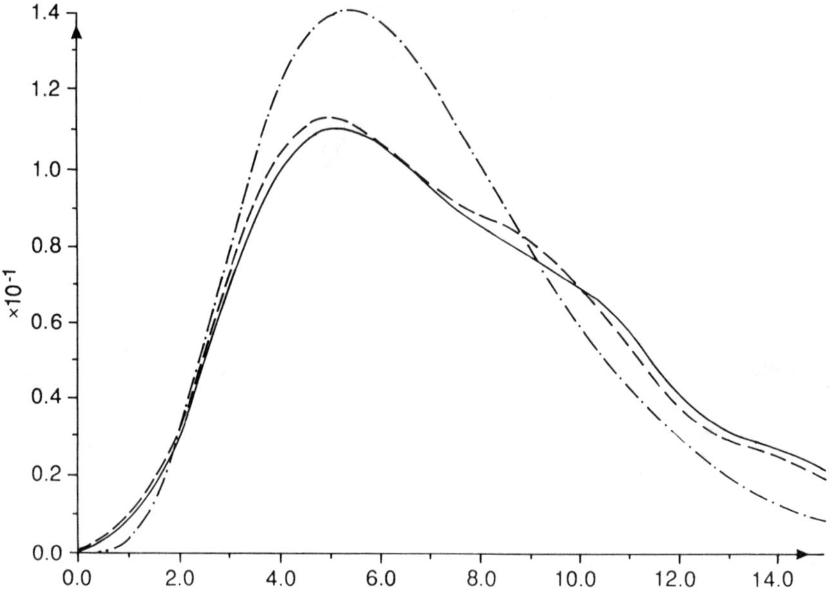

Figure 1c. *Kernel density estimates for a* χ^2 *test statistic based on 1000 pseudo samples* (———) *and based on one bootstrap estimate with 1000 replications* (— — —) *and the asymptotic* χ^2 *density (with 10 degrees of freedom, up to a factor* (— · — · —) *(Model A)*

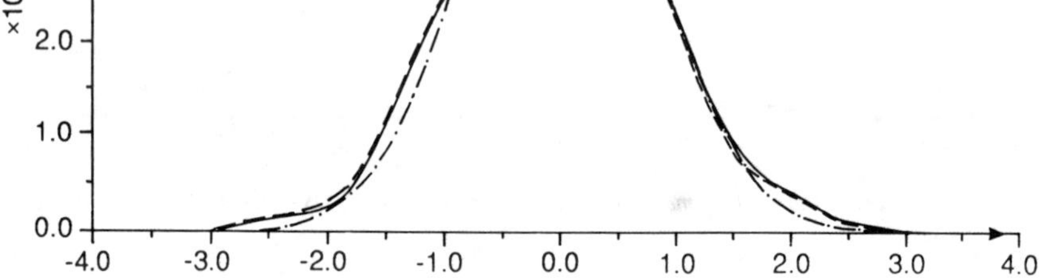

Figure 2a. *Kernel density estimates for the linear contrast* $c_1^T(\hat{\beta} - \beta)$ *based on 1000 pseudo samples* (———) *and based on one bootstrap estimate with 1000 replications* (— — —) *and the asymptotic* χ^2 *density (with 10 degrees of freedom, up to a factor* (— · — · —) *(Model B)*

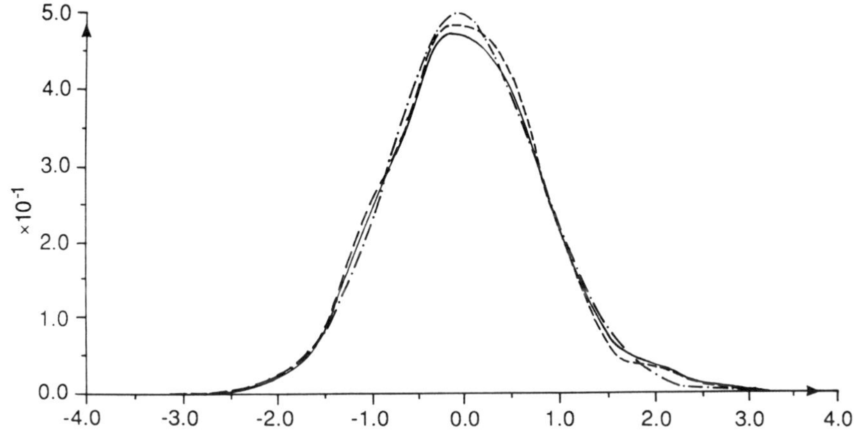

Figure 2b. *Kernel density estimates for the linear contrast* $c_2^T(\hat{\beta} - \beta)$ *based on 1000 pseudo samples (———) and based on one bootstrap estimate with 1000 replications (— — —) and the asymptotic* χ^2 *density (with 10 degrees of freedom, up to a factor (—·—·—) (Model B)*

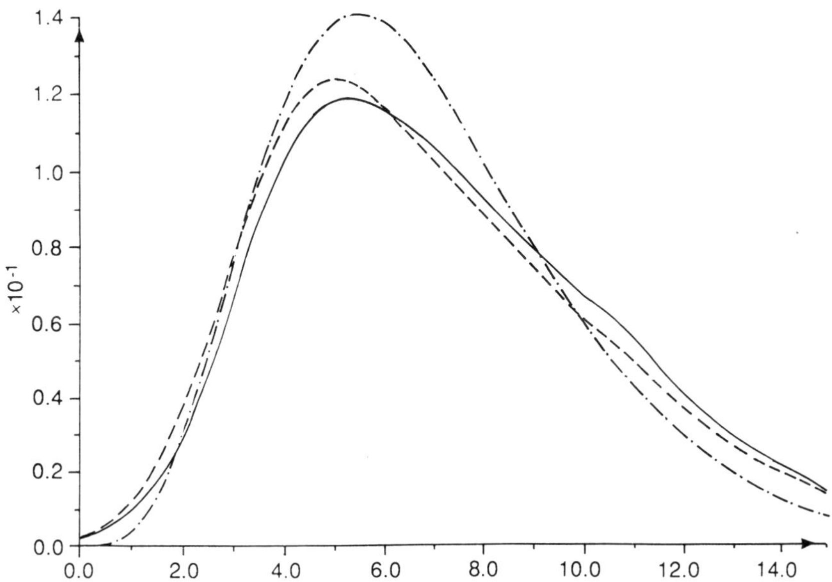

Figure 2c. *Kernel density estimates for a* χ^2 *test statistic based on 1000 pseudo samples (———) and based on one bootstrap estimate with 1000 replications (— — —) and the asymptotic* χ^2 *density (with 10 degrees of freedom, up to a factor (—·—·—) (Model B)*

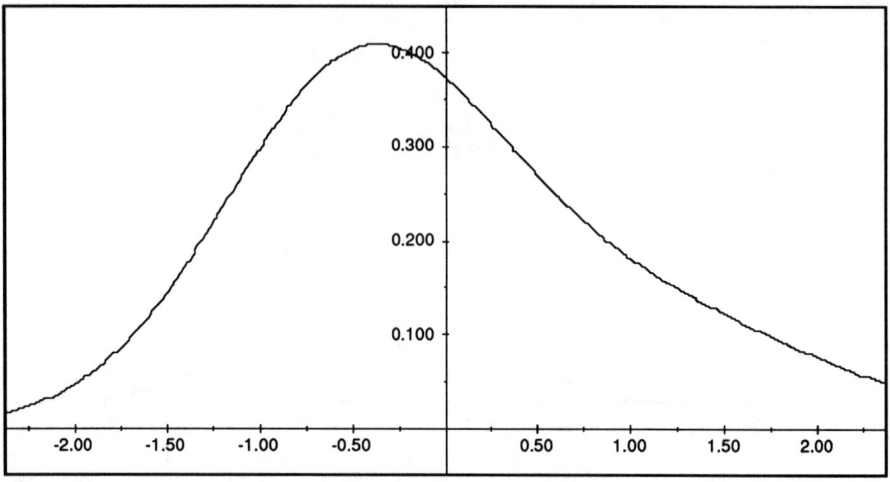

Figure 3. *Density of the error variables* ε_i *in the simulation of Section 2.*

Chapter 8. Bootstrap and wild bootstrap for high - dimensional linear random design models.

1. Introduction. In this chapter two bootstrap procedures are studied for high - dimensional linear models where the design vectors are random and the dimension p of the parameter is large. The stochastic structure of this model is more complex than in the case of linear models with nonrandom design which have been considered in the last two chapters. There the bootstrap procedure has to mimic the relatively simple case of one dimensional i. i. d. error variables. Now the model will be described by distributions in high dimensional spaces. We will see that bootstrap works in this more complex model not always as satisfactorily as in the case of a nonrandom design.

We will consider linear contrasts and F - test statistics. As in Chapters 6 and 7 an asymptotic approach will be chosen where the dimension p of the model may increase for sample size $n \to \infty$. The range of validity will be compared for the normal approximation and for the bootstrap procedures. Furthermore it will be argued that the rates of convergence are different for the bootstrap procedures in this asymptotic framework. This is in contrast to the usual asymptotic approach where p is fixed.

We consider the linear model

(1.1) $$Y_i = X_i^T\beta + \varepsilon_i \quad (i=1,\ldots,n).$$

In this equation Y_i are n observations, X_i are the (observed) random design vectors belonging to $\mathbf{R}^p$, β is a p-dimensional parameter, and ε_i are the (unobservable) error variables. The stochastic structure is described by

(1.2) (X_i,Y_i) *are i. i. d. with finite second moments* $E\, Y_i^2 < \infty$ *and* $E\, \|X_i\|^2 < \infty$,

(1.3) β *minimizes* $b \to E(Y_i - X_i^T b)^2$.

We are interested in the estimation of the distribution of the least squares estimator $\widehat{\beta}$ of β

(1.4) $$\widehat{\beta} = (\mathbf{X}^T\mathbf{X})^{-1}\mathbf{X}^T\mathbf{Y}$$

where $\mathbf{X}$ is the n×p data matrix with rows X_i^T and $\mathbf{Y}$ is an n×1 data vector with elements Y_i.

Note that (1.2) and (1.3) imply that $E\, X_i\, \varepsilon_i = E\, X_i\, (Y_i - X_i^T\beta) = 0$ holds. But these assumptions do *not* imply that

(1.5) *given* $(X_1,...,X_n)$ *the* ε_i $(i=1,...,n)$ *are conditionally i. i. d. with mean zero.*

If assumption (1.5) held one could apply the bootstrap which is based on resampling from the centered residuals $\hat{\varepsilon}_i - \hat{\varepsilon}_{\cdot}$. Here $\hat{\varepsilon}_i = Y_i - X_i^T\hat{\beta}$ are the residuals and $\hat{\varepsilon}_{\cdot} = (1/n)\sum_{i=1}^n \hat{\varepsilon}_i$. This bootstrap procedure has been discussed in length in Chapter 6.

In this chapter we treat the case where the assumption of (1.5) is not appropriate. Take for instance the case where the conditional distributions of the errror variables ε_i are heteroschedastic. Then the bootstrap described above is not consistent with the model. Note also that the conditions (1.2) and (1.3) do not imply that the linear model holds in the sense that $E(\,\varepsilon_i \mid X_i = x\,)$ vanishes for every x. The parameter β defines the "nearest" linear model. For instance this model may be appropriate if we are interested in the best approximation of $x \to E(\,Y_i \mid X_i = x\,)$ in a given finite dimensional linear function space (for instance polynomial regression). Also in the usual setup where one assumes that $E(\,\varepsilon_i \mid X_i\,) = 0$ holds it may be interesting how bootstrap behaves under violation of this assumption. Therefore we will in general not assume that $E(\,\varepsilon_i \mid X_i\,) = 0$ holds.

In this chapter we study a bootstrap and a wild bootstrap procedure for the model (1.2), (1.3). The following bootstrap procedure is due to Efron (1982).

BOOTSTRAP. *One generates a resample* $(X_1^*,Y_1^*),...,(X_n^*,Y_n^*)$ *from* $\{(X_1,Y_1), ..., (X_n,Y_n)\}$. *Then one defines*

$$\hat{\beta}^* = (\mathbf{X}^{*T}\mathbf{X}^*)^{-1}\, \mathbf{X}^{*T}\mathbf{Y}^* \tag{1.6}$$

and one uses $\mathcal{L}^*(\sqrt{n}(\hat{\beta}^* - \hat{\beta}))$ *as an estimate of* $\mathcal{L}(\sqrt{n}(\hat{\beta}-\beta))$ *where* $\mathcal{L}^*(...)$ *denotes the conditional distribution* $\mathcal{L}(\ \mid X_1,...,X_n,Y_1,...,Y_n)$.

The consistency of this procedure has been proved in Freedman (1981) for the case of fixed dimension p. As second possibility for resampling we consider in this chapter wild bootstrap (see Chapters 1 and 2). Wild bootstrap has been proposed for the case that the additional assumption $E(\varepsilon_i|X_i) = 0$ is appropriate. Nevertheless we will study this procedure also for the case that this assumption is violated. Here it proceeds as follows.

WILD BOOTSTRAP. *First one estimates* $\mathcal{L}(\varepsilon_i|X_1,...,X_n)$ *by an arbitrary distribution* $\widehat{F}_i$ *with*

(1.7) $$E(Z\,|\,\widehat{F}_i) = 0,$$

(1.8) $$E(Z^2\,|\,\widehat{F}_i) = \hat{\varepsilon}_i^2,$$

(1.9) $$E(Z^3\,|\,\widehat{F}_i) = \hat{\varepsilon}_i^3.$$

Then one generates independent ε_i^W *with* $\mathcal{L}(\varepsilon_i^W) = \widehat{F}_i$ *and one lets*

(1.10) $$Y_i^W = X_i^T\hat{\beta} + \varepsilon_i^W \quad (i=1,...,n).$$

$\hat{\beta}^W$ *is the least squares estimator based on the resample*

(1.11) $$\hat{\beta}^W = (X^TX)^{-1}X^TY^W.$$

As we have mentioned in Chapter 1 this bootstrap procedure has been called *wild* bootstrap because n (!) different distributions $\mathcal{L}(\varepsilon_i|X_i)$ are estimated by only n observations. The condition (1.9) has been introduced to improve the rate of convergence of the bootstrap estimate. In this chapter we will always assume that one of the three constructions of $\widehat{F}_i$ is used which has been proposed in the discussion after Theorem 4 in Chapter 1. In the simulations we choose the third construction, i. e. $\varepsilon_i^W = ((\delta_1 + V_{i,1}/\sqrt{2})(\delta_2 + V_{i,2}/\sqrt{2}) - \delta_1\delta_2)\,\hat{\varepsilon}_i$ where the $V_{i,j}$'s are independent N(0,1) - distributed variables and where $\delta_1 = (3/4 + \sqrt{17}/12)^{1/2}$ and $\delta_2 = (3/4 - \sqrt{17}/12)^{1/2}$.

We will use an asymptotic approach where the dimension p of the random design vectors X_i may increase rapidly with the sample size n but where assumptions are made on (the tails of) the distributions of X_i which ensure that the random p×p matrix $\frac{1}{n}\sum_{i=1}^n X_iX_i^T$ is invertible with high probability and that $(\frac{1}{n}\sum_{i=1}^n X_iX_i^T)^{-1} \approx (EX_iX_i^T)^{-1}$. The conditions will exclude the interesting case that some observations become very influential compared with the other observations (leverage points). We have not been able to state asymptotic results in an asymptotic framework which admits increasing dimension p *and* the occurrence of leverage points.

This chapter is organized as follows. In Section 2 we show that for linear contrasts bootstrap and wild bootstrap work for high - dimensional models in the whole range of the validity of asymptotic normality. In the third section the accuracy of these bootstrap procedures is compared with the normal approximation. This comparison will be done by formal Edgeworth expansions and simulations. In Section 4 we show consistency of bootstrapping a F - test statistic under conditions which in general do not imply the validity of the approximation by a F - distribution. The proofs of the theorems are given in Section 5.

2. Consistency of bootstrap for linear contrasts. In this section the bootstrap of the distribution of linear contrasts $c^T(\hat{\beta} - \beta)$ will be considered. It will turn out (see Theorem 1) that bootstrap estimates consistently the unconditonal law $\mathcal{L}(\sqrt{n}\, c^T(\hat{\beta} - \beta))$ under weaker assumptions on (the dimension of) the model than those which are necessary for wild bootstrap and normal approximations.

Because in our approach everything may depend on n, dependence on n is not indicated in the notation. The following conditions are always fulfilled after a standardization of the observations (X_i, Y_i) and of the parameter β.

(2.1) $$E\, X_i X_i^T = I_p,$$

(2.2) $$0 < \inf_n E\, \varepsilon_i^2 \le \sup_n E\, \varepsilon_i^2 < \infty.$$

For instance to get (2.1) define X_i as $(E\, X_i X_i^T)^{-1/2} X_i$ and β as $(E\, X_i X_i^T)^{1/2} \beta$. Note again that this corresponds to the usual $n^{1/2}$ renorming of the parameter.

As in the introduction we denote $\mathcal{L}(\ldots | X_1,\ldots,X_n,Y_1,\ldots,Y_n)$ by $\mathcal{L}^*(\ldots)$. d_∞ is the Kolmogorov distance (sup norm between the distribution functions).

THEOREM 1. *Choose* $c \in \mathbf{R}^p$ *with* $\|c\| = 1$. *Consider a linear model* (1.1), ..., (1.3) *which is standardized such that* (2.1) *and* (2.2) *hold. For a* $\delta > 0$ *assume that*

(2.3) $$p^{1+\delta}/n \to 0,$$

(2.4) $\sup_{n\geq 1} \sup_{\|d\|=1} E\, |d^T X_i|^{4K} (1+\varepsilon_i^2) < \infty$, *where* K *is the smallest integer* $\geq 2/\delta$,

(2.5) $E\, (c^T X_i)^2 \varepsilon_i^2\, I[\, (c^T X_i)^2 \varepsilon_i^2 \geq \gamma\, n\,] \to 0$ *for every* $\gamma > 0$.

Then

(2.6) $d_\infty(\mathcal{L}^*(\sqrt{n}\, c^T (\hat\beta^* - \hat\beta)), \mathcal{L}(\sqrt{n}\, c^T(\hat\beta - \beta))) \to 0$ *(in probability), if* $E(\varepsilon_i|X_i) = 0$ *or if* $\delta \geq 1/3$,

(2.7) $d_\infty(\mathcal{L}^*(\sqrt{n}\, c^T(\hat\beta^W - \hat\beta)), \mathcal{L}(\sqrt{n}\, c^T(\hat\beta - \beta))) \to 0$ *(in probability), if* $E(\varepsilon_i|X_i) = 0$ *or if* $\delta \geq 1$.

Note that for decreasing δ condition (2.3) becomes weaker and condition (2.4) becomes stronger. We conjecture that the moment conditions in (2.4) may be replaced by weaker but more complicated conditions on the distribution tails of the design vectors. For instance for the case that the elements of X_i are i. i. d. the results of Yin et.al.(1988) suggest that it suffices that the fourth moments of the elements of X_i are bounded.

Given the conditions (2.3), (2.4) the Lindeberg condition (2.5) is necessary for the asymptotic normality of $\sqrt{n}\, c^T(\hat\beta - E\, \hat\beta)$. This follows from Lemma 3 in Section 5. Therefore Theorem 1 shows that, given (2.3), (2.4), bootstrap and wild bootstrap work in the whole range of the validity of asymptotic normality of $\sqrt{n}\, c^T(\hat\beta - E\, \hat\beta)$.

If $E(\varepsilon_i|X_i) \neq 0$ in general the bias $\sqrt{n} c^T(E\, \hat\beta - \beta)$ is of order $p/\sqrt{n}$ which may tend to infinity if $\delta < 1$. Then for $\sqrt{n}\, c^T(\hat\beta - \beta)$ a mean zero normal approximation and wild bootstrap fail whereas bootstrap works as long as $\delta \geq 1/3$ (i.e. $p^4/n^3 \to 0$). Therefore bootstrap works under weaker model assumptions than wild bootstrap (which is natural by the construction of these procedures). But in Section 3 we will see that this model robustness of bootstrap must be paid for by a slower rate of convergence than wild bootstrap (for the estimation of the studentized estimator) if the linear model holds i. e. $E(\varepsilon_i|X_i) = 0$.

Under the assumptions of Theorem 1 bootstrap and wild bootstrap estimates also consistently the conditional law $\mathcal{L}(\sqrt{n}\, c^T (\hat{\beta} - \beta) \mid X_1,\dots,X_n)$ if additionally $E(\varepsilon_i \mid X_i) = 0$ holds or if $\delta \geq 1$ (i. e. $p^2 / n \to 0$). This holds because under these assumptions $d_\infty(\mathcal{L}(\sqrt{n} c^T (\hat{\beta} - \beta) \mid X_1,\dots,X_n), \mathcal{L}(\sqrt{n}\, c^T (\hat{\beta} - \beta))) \to 0$ (in probability). To see this note that the conditonal and the unconditional law converge to the same normal limit $N(0,E(c^T X_i)^2 \varepsilon_i^2)$. For the unconditional law this follows from Lemma 2 and 4 in Section 5. For the conditional law one argues similarly as in the proofs of Lemma 4* and 6* in Section 5.

If $E(\varepsilon_i \mid X_i) = 0$ then for the study of the wild bootstrap another approach than outlined in Theorem 1 and in its proof may be more appropriate. Consider the case that the conditional distribution of $c^T\hat{\beta}$ given a sequence of realizations of $(X_1,\dots,X_n)$ can be approximated by a normal distribution. This would imply the Lindeberg condition. Then the Lindeberg condition holds also for $c^T(\hat{\beta}^W - \hat{\beta})$ (see Lemma 6*) and under weak additional assumptions the same normal approximation works also for the conditional law of $c^T(\hat{\beta}^W - \hat{\beta})$. This would entail that wild bootstrap approximates $\mathcal{L}(\sqrt{n}\, c^T(\hat{\beta} - \beta) \mid X_1,\dots,X_n)$ consistently nearly as long as a mean zero normal approximation works.

3.Accuracy of the bootstrap. In this section the cases will be studied more closely where bootstrap and normal approximations work. We conjecture that for high dimensional models wild bootstrap is a more accurate estimate of the distribution of $\sqrt{n}\; c^T(\hat{\beta} - \beta)$ than bootstrap if the bias $\sqrt{n}\, c^T(E\,\hat{\beta} - \beta)$ is not too large. The accuracy of bootstrap and normal approximations will be compared by a formal Edgeworth expansion and by simulations. For deriving the formal Edgeworth expansion we assume that the linear model holds in the sense of $E(\varepsilon_i \mid X_i) = 0$. Then $E\,\hat{\beta} = \beta$. We will give no proof of the validity of the Edgeworth expansion. We will show the rates of convergence only for the bootstrap estimates of the first four moments of a linear contrast $c^T(\hat{\beta} - \beta)$ and we will use these results for an heuristic study of bootstrap based on the assumption that the formal Edgeworth expansion holds: $P(Z \leq x) = \Phi(\tilde{x}) + \gamma(Z)\,(1-\tilde{x}^2)\,\varphi(\tilde{x}) + O(\delta(Z))$ where $\tilde{x} = (x - EZ)\, var(Z)^{-1/2}$, $\gamma(Z) = (1/6)\, E\,\tilde{Z}^3$ and $\delta(Z) = E\,\tilde{Z}^4 - 3 + \gamma(Z)^2$ with $\tilde{Z} = (Z - EZ)\, var(Z)^{-1/2}$. A higher order analysis for wild bootstrap based on Edgeworth expansions can be found in Liu(1988) for a one dimensional linear model and in Cao - Abad (1990) and Cao - Abad and Gonzales - Manteiga (1990) for nonparametric regression. For a simple approach of a higher order analysis of bootstrap which does not use Edgeworth expansions see Chapter 5.

For the analysis of bootstrap we consider the moments of the following expansions of $\sqrt{n}\ c^T(\hat{\beta} - \beta)$ and $\sqrt{n}\ c^T(\hat{\beta}^* - \hat{\beta})$:

$$(3.1)\qquad \hat{\theta}_{c,n} = \sum_{h=0}^{2} c^T A^h \frac{1}{\sqrt{n}} \sum_{i=1}^{n} X_i \varepsilon_i,$$

$$(3.2)\qquad \hat{\theta}^*_{c,n} = \sum_{h=0}^{2} c^T A^{*h} \frac{1}{\sqrt{n}} \sum_{i=1}^{n} X_i^* \varepsilon_i^*,$$

where $A = I - 1/n \sum_{i=1}^{n} X_i X_i^T$ and $A^* = I - 1/n \sum_{i=1}^{n} X_i^* (X_i^*)^T$.

THEOREM 2. *Choose* $c \in \mathbf{R}^p$ *with* $\|c\|=1$. *Consider a linear model with* (1.1) ,..., (1.3), *which is standardized such that* (2.1) *and* (2.2) *hold. Assume additionally* $E(\varepsilon_i|X_i) = 0$ *and*

$$(3.3)\qquad p^{1+\delta}/n \to 0 \textit{ for a } \delta > 0.$$

If the following holds for large enough C_1, C_2, *and* $C_3 > 0$ *(depending on* δ *)*

$$(3.4)\qquad \sup_n E\,(\|X_i\| / \sqrt{p})^{C_1} < +\infty,$$

$$(3.5)\qquad \sup_n \sup_{\|d\|=1} E\,(d^T X_i)^{C_2} < +\infty,$$

$$(3.6)\qquad \sup_n E\,|\varepsilon_i|^{C_3} < +\infty$$

then

$$(3.7)\qquad \hat{\theta}_{c,n} = \sqrt{n}\, c^T(\hat{\beta} - \beta) + o_P(p/n),$$

$$(3.8)\qquad \hat{\theta}^*_{c,n} = \sqrt{n}\, c^T(\hat{\beta}^* - \hat{\beta}) + o_P(p/n).$$

For the moments of the stochastic expansions $\hat{\theta}_{c,n}$ *and* $\hat{\theta}^*_{c,n}$ *one gets*

$$(3.9)\qquad E^* \hat{\theta}^*_{c,n} = O_P(p/n),$$

$$(3.10)\qquad \mathrm{var}^*(\hat{\theta}^*_{c,n}) - \mathrm{var}(\hat{\theta}_{c,n}) = O_P(n^{-1/2} + p/n),$$

(3.11) $\qquad \mu_3^*(\widehat{\theta}^*_{c,n}) - \mu_3(\widehat{\theta}_{c,n}) = O_P(n^{-1} + p/n^{3/2}),$

(3.12) $\qquad \mu_4^*(\widehat{\theta}^*_{c,n}) - 3\,\mathrm{var}^*(\widehat{\theta}^*_{c,n})^2 = O_P(n^{-1}),$

(3.13) $\qquad \mu_4(\widehat{\theta}_{c,n}) - 3\,\mathrm{var}(\widehat{\theta}_{c,n})^2 = O(n^{-1}).$

Here for $j = 3$ *and* $j = 4$ *the quantities* μ_j, μ_j^*, *and* var^* *are defined as* $\mu_j(Z) = E((Z - E(Z))^j)$, $\mu_j^*(Z) = E^*((Z - E^*(Z))^j)$, *and* $\mathrm{var}^*(Z) = E^*(Z - E^*(Z))^2$, *where* $E^*(Z) = E(Z \mid X_1, \ldots, X_n, Y_1, \ldots, Y_n)$.

Theorem 2 can be used for an heuristic study of the accuracy of the bootstrap. (3.10) suggests that the sup norm between the distribution function of $\sqrt{n}\; c^T(\widehat{\beta} - \beta)$ and the bootstrap approximation is of order $O_P(n^{-1/2} + p/n)$. This rate of convergence may be very slow for large p. For many cases it has been proposed to bootstrap the studentized estimator to increase the rate of convergence (see Chapers 1 and 5). But (3.9) suggests that the rate of convergence of the bootstrap approximation of the studentized linear contrast is of order $O_P(p/n)$. Note that here for large p ($p > \sqrt{n}$) studentization does not increase the rate of convergence.

The situation is quite different for wild bootstrap. Conditionally on $(X_1, \ldots, X_n, Y_1 \ldots, Y_n)$ the estimate $\sqrt{n} c^T(\widehat{\beta}^W - \widehat{\beta})$ is a sum of conditionally independent zero mean variables. This simplifies the calculation of moments (and would also make a rigorous proof of the Edgeworth expansion easier). Because the moments of the statistic $\sqrt{n}\; c^T(\widehat{\beta} - \beta)$ may not exist under our assumptions we consider conditional moments given the following event $H = H_\varepsilon = \{$The maximal absolute eigenvalue of $\mathbf{X}^T\mathbf{X}/n - I$ is bounded by $(p/n)^\varepsilon\}$ for an ε with $0 < \varepsilon < 1/2$.

THEOREM 3. *Choose* $\kappa > 0$ *and* ε *with* $0 < \varepsilon < 1/2$. *Then under the assumptions of Theorem 2 with appropriately chosen* C_1, C_2, *and* $C_3 > 0$ *one gets*

(3.14) $\qquad P(H_\varepsilon) = 1 - O(n^{-\kappa}),$

(3.15) $\qquad \mathrm{var}_H^*(n^{1/2} c^T(\widehat{\beta}^W - \widehat{\beta})) - \mathrm{var}_H(n^{1/2} c^T(\widehat{\beta} - \beta)) = O_P(n^{-1/2} + p/n),$

(3.16) $\qquad \mu_{3,H}^*(n^{1/2} c^T(\widehat{\beta}^W - \widehat{\beta})) - \mu_{3,H}(n^{1/2} c^T(\widehat{\beta} - \beta)) = O_P(n^{-1} + p\, n^{-3/2}),$

$$(3.17) \qquad \mu_{3,H}(\, n^{1/2}\, c^T(\widehat{\beta} - \beta)\,) = O_P(n^{-1/2}),$$

$$(3.18) \qquad \mu^*_{4,H}(n^{1/2}\, c^T(\widehat{\beta}^W - \widehat{\beta})) - 3\, \mathrm{var}^*_H(n^{1/2}\, c^T(\widehat{\beta}^W - \widehat{\beta}))^2 = O_P(n^{-1}),$$

$$(3.19) \qquad \mu_{4,H}(n^{1/2}\, c^T(\widehat{\beta} - \beta)) - 3\, \mathrm{var}_H(n^{1/2}\, c^T(\widehat{\beta} - \beta))^2 = O_P(n^{-1}).$$

Here for $j = 3$ *and* $j = 4$ *the quantities* $\mu_{j,H}$, $\mu^*_{j,H}$, var_H *and* var^*_H *are defined as* $\mu_{j,H}(Z) = E(\,(Z - E(Z|H))^j\, |H)$, $\mu^*_{j,H}(Z) = E^*_H(\,(Z - E^*_H(Z))^j)$, $\mathrm{var}_H(Z) = E(\,(Z - E(Z|H))^2\, |H)$ *and* $\mathrm{var}^*_H(Z) = E^*_H(Z - E^*_H(Z))^2$, *where* $E^*_H(Z) = E(Z \mid H, X_1, \ldots, X_n, Y_1, \ldots, Y_n)$ *and* $H = H_\varepsilon$.

Theorem 3 suggests that wild bootstrap of an unstudentized linear contrast has the same rate of convergence as bootstrap. However in contrast to bootstrap, studentization leads always to an improvement of the rate of convergence of wild bootstrap. Then wild bootstrap produces approximations of order $O_P(\, n^{-1} + p\, n^{-3/2})$. For studentization one may use the variance estimate $\widehat{\sigma}^2_c = \mathrm{var}^W(\sqrt{n} c^T \widehat{\beta}^W) = 1/n \sum_{i=1}^n (c^T(X^TX/n)^{-1}X_i)^2(Y_i - X_i^T\widehat{\beta})^2$. Because of (3.15) the accuracy of the normal approximation $N(0, \widehat{\sigma}^2_c)$ is $O_p(\, n^{-1/2} + p\, n^{-1})$. The different rates of convergence are summarized in Table 1.

Estimation of	$\mathcal{L}(\sqrt{n}\, c^T(\widehat{\beta} - \beta)\,)$	$\mathcal{L}(\sqrt{n}\, c^T(\widehat{\beta} - \beta) / \widehat{\sigma}_c\,)$
Normal approximation $N(0,\widehat{\sigma}^2_c)$	$O_P(\, n^{-1/2} + p\, n^{-1})$	-
Wild bootstrap	$O_P(\, n^{-1/2} + p\, n^{-1})$	$O_P(\, n^{-1} + p\, n^{-3/2})$
Bootstrap	$O_P(\, n^{-1/2} + p\, n^{-1})$	$O_P(\, p\, n^{-1})$

Table 1. *Rates of convergence of the bootstrap procedures and the mean zero normal approximation under the assumption* $E(\varepsilon_i \mid X_i) = 0$.

We have also compared bootstrap, wild bootstrap, and normal approximation in a simulation study. The results are displayed in Figures 1aa to 4d. We have considered the following model. The sample size n is 50 and the dimension p is 5 or 10. The design vectors are defined as

$$X_{i1} = 1 \text{ and } X_{ij} = U_{ij}\, Z_i/2 \qquad (i = 1, \ldots, 50;\, j = 2, \ldots, p)$$

and for different choices of $\beta = (\beta_1, \beta_2, \beta_3)$ the observations are put as

$$Y_i = \beta_1 Q_i + \beta_2 Q_i V_i + \beta_3 V_i,$$

where $Q_i = (\| X_i \|^2 - 1 - (p-1)/4 E Z_i^2)/(p-1)$.

The variables $U_{1,2}, \dots, U_{50,p}, V_1, \dots, V_{50}, Z_1, \dots, Z_{50}$ are independent with the following distributions. The $U_{i,j}$'s have a standard normal distribution N(0,1). The distribution of the V_i's is a mixture of normal distributions: $1/2\, N(1/2, (1.2)^2) + 1/2\, N(-1/2, (0.7)^2)$. This error distribution has also been used in the simulation of Chapter 7. There also a plot of the density of this distribution has been given (see Figure 3 in Chapter 7). The Z_i's are uniformly distributed on the interval [1,3]. We have used the following choices of β:

$$\beta = (0, 0, 1) \qquad \text{(CASE A)},$$

$$\beta = (0, 4, 1) \qquad \text{(CASE B)},$$

$$\beta = (4, 1, 0) \qquad \text{(CASE C)}.$$

In case A the error variables ε_i are conditionally i. i. d. (given $X_1, \dots, X_n$) with $E(\varepsilon_i \mid X_i) = 0$. In case B the linear model also holds in the sense of $E(\varepsilon_i \mid X_i) = 0$ but the errror variables ε_i are conditionally heteroschedastic. In case C the conditional expectation $E(\varepsilon_i \mid X_i)$ does not vanish. The Monte Carlo calculations are designed as in Chapter 1. However, we have used now a smaller number of simulations. First we used 1000 simulations to calculate the .01, j/20 (for j =1, ..., 19), and .99 quantiles of the distribution of linear contrasts $c^T(\hat{\beta} - \beta)$ and of studentized linear contrasts $c^T(\hat{\beta} - \beta) / \hat{\sigma}$. This has been done for two linear contrasts: $c = c_1 = (1, 0, \dots, 0)^T$ and $c = c_2 = (0, 1, 0, \dots, 0)^T$. The bootstrap procedure, the wild bootstrap procedure, and the normal approximation (with estimated variance) for estimating the distribution function of these two linear contrasts (studentized and unstudentized) have been calculated. Every bootstrap resampling and every wild bootstrap resampling uses 1000 replications. In the Figures the Monte Carlo estimate for the expected squared error of the bootstrap procedures and the normal approximation are plotted against the distribution function of the linear contrasts $c^T(\hat{\beta} - \beta)$ or of the studentized linear contrasts $c^T(\hat{\beta} - \beta) / \hat{\sigma}$. For p = 5 also the curves of bias and standard deviation have been plotted. In Tables 2, ..., 4 expectations and standard deviations of the procedures are given at the 10% and the 90% quantiles.Taking into account that we have now used a smaller number of simulations than in Chapter 1 we expect here a standard deviation of the Monte Carlo calculations of less than 8% of the plotted expected squared errors.

We draw the following conclusions from our simulations.

♣ Consider first the case of the first linear contrast c_1 in case C (β = (4, 1, 0). In this case $c_1^T(\widehat{\beta}-\beta)$ has a large bias: $E\, c_1^T(\widehat{\beta}-\beta) / (\, var(c_1^T\widehat{\beta})^{1/2}$ = - 0.77 (for p = 5) and = - 0.58 (for p = 10). The bootstrap and wild bootstrap estimate (and the mean zero normal approximation) of the distribution function are shifted compared with the real distribution function. The bootstrap estimate gives here the best approximation. This is in accordance with Theorem 1 which shows that bootstrap is more robust against departures of $E(\varepsilon_i \mid X_i)$ from 0.

♣ The situation is quite different for the second linear contrast c_2. $c_2^T(\widehat{\beta}-\beta)$ has only a small bias $E\, c_2^T(\widehat{\beta}-\beta) / (\, var(c_2^T\widehat{\beta})^{1/2}$ = - 0.009 (for p = 5) and = - 0.017 (for p = 10). Here one has a similar picture as in cases A and B. The location of the real distribution is estimated more accurately by all estimates. The bootstrap estimate has a large variance compared with the wild bootstrap estimate and with the estimate based on the normal approximation. The wild bootstrap estimate of the studentized linear contrast produces here the best approximation.

♣ Consider again cases A and B and the case of the second linear contrast c_2 in case C. The wild bootstrap estimate works here *much* better for the studentized linear contrast than for the unstudentized contrast. This is *not* always true for the bootstrap estimate (in particular, see the tails of the distributions in Figures 2aa, 2ad, 2ba, 2bd,3ba). Also in other simulations (where we choose β = (8, 0, 0) and where the V_i's have a standard normal distribution (CASE D)) bootstrap of the unstudentized estimate works slightly better in the tails than bootstrap of the studentized estimate (see Figures 4a and 4d). Note that this contradicts usual higher - order results for fixed dimension p which say that studentization improves the rate of convergence of bootstrap (see Chapter 5). But this is in accordance with our discussion based on Edgeworth expansions where we have argued that studentization of the linear contrast does not lead to a faster rate of convergence of the bootstrap estimate if the dimension p is very large. (In case D the values of u and v have been chosen in advance and have been not approximated using simulations. At the chosen points u and v the distribution functions of the linear contrasts have been calculated by simulations. In Figures 4a, ..., 4d these distribution functions are plotted against the expected squared error, bias and standard deviation of bootstrap. Again, every marked point corresponds to one simulation results. The marke points are now not equidistant.)

♣ Note that also in case C for $c = c_2$ wild bootstrap is preferable to bootstrap although in this model the departure from the linear model is drastic: $E(\varepsilon_i \mid X_i)$ is of the same order as the conditional standard deviation $(\mathrm{var}\,(\varepsilon_i \mid X_i))^{1/2}$ of the error variable.

4. Bootstrapping F-tests. Bootstrap may be used to estimate critical values of a F-test statistic. In this section we will show that bootstrap works in high - dimensional linear models under weak conditions (see Theorem 3). Note that in our model (1.1) ,..., (1.3) in general the approximation by an F distribution fails.

For linear subspaces $\mathbf{H}_0 \subset \mathbf{H}_1$ of $\mathbf{R}^p$ $(0 \in \mathbf{H}_0)$ we consider the testing problem $\beta \in \mathbf{H}_0$ versus $\beta \in \mathbf{H}_1$. The F-test statistic for this testing problem is defined by

$$T = \frac{\|\Pi_1 \mathbf{Y} - \Pi_0 \mathbf{Y}\|^2 / (p_1 - p_0)}{\|\mathbf{Y} - \Pi_1 \mathbf{Y}\|^2 / (n - p_1)}$$

where (for i=1,2) $\Pi_i \mathbf{Y}$ is the projection of $\mathbf{Y}$ onto $L_i = \{z \mid z_j = X_j^T b$ for a $b \in \mathbf{H}_i\}$ and p_i is the dimension of L_i. Denote the least squares estimator under the hypothesis $\beta \in \mathbf{H}_i$ by $\hat{\beta}_i$. For the determination of critical values we want to estimate the distribution of T for parameters β in the hypothesis $\mathbf{H}_0$. Furthermore on the alternative $\beta \in \mathbf{H}_1 \backslash \mathbf{H}_0$ the bootstrap estimate of the distribution of T should converge to the distribution of T under a parameter β_0 which lies in the hypothesis $\mathbf{H}_0$. This would guarantee a satisfactory power of T when the bootstrap critical values are used.

In the next theorem we will show that bootstrap and wild bootstrap work after the following slight modifications.

(4.1) $\{(X_i^*, Y_i^*)\}$ *is a resample from* $\{(X_i, Y_i - X_i^T\hat{\beta}_1 + X_i^T\hat{\beta}_0)\}$

(4.2) *The conditional variance and the conditional third moment of* ε_i^W *are* $(Y_i - X_i^T\hat{\beta}_1)^2$ *and* $(Y_i - X_i^T\hat{\beta}_1)^3$ *(respectively). The* Y_i^W *are defined as* $Y_i^W = X_i^T\hat{\beta}_0 + \varepsilon_i^W$.

Using these bootstrap observations one can construct bootstrap test statistics

$$T^* = \frac{\|(\Pi_1-\Pi_0)(\mathbf{Y}^*)\|^2/(p_1-p_0)}{\|(I-\Pi_1)(\mathbf{Y}^*)\|^2/(n-p_1)},$$

and

$$T^W = \frac{\|(\Pi_1-\Pi_0)(\mathbf{Y}^W)\|^2/(p_1-p_0)}{\|(I-\Pi_1)(\mathbf{Y}^W)\|^2/(n-p_1)}.$$

THEOREM 4. *Assume for a* $\beta \in \mathbf{H}_1$ (1.1),(1.2) *with* $E(\varepsilon_i|X_i)=0$, *and suppose that the model is standardized such that* (2.1) *and* (2.2) *hold. Assume that for a* $\delta > 0$

(4.2) $\quad p_1^{1+\delta}/n \to 0,$

(4.3) $\quad \sup_{n\geq 1} \sup_{\|d\|=1, d\in H_1} E(d^T X_i)^{4K}(1+\varepsilon_i^2) < \infty$, *where* K *is the smallest integer* $\geq 2/\delta$,

(4.4) $\quad p_1 \sqrt{p_1 - p_0}\,/\,n \to 0,$

(4.5) $\quad \sup_{n\geq 1} \sup_d E\,(d^T X_i)^4 \varepsilon_i^4 < \infty$, *where the second supremum is taken over* $d \in \mathbf{H}_1 \cap \mathbf{H}_0^\perp$ *with* $\|d\| = 1$.

Then the following holds for the resampling plans (4.1) *and* (4.2)

(4.6) $\quad d_\infty(\mathcal{L}^*(T^*), \mathcal{L}(T^0)) \to 0$ *(in probability),*

(4.7) $\quad d_\infty(\mathcal{L}^*(T^W), \mathcal{L}(T^0)) \to 0$ *(in probability),*

where

$$T^0 = \frac{\|(\Pi_1-\Pi_0)(\mathbf{Y}-\mathbf{X}\beta)\|^2/(p_1-p_0)}{\|(I-\Pi_1)(\mathbf{Y})\|^2/(n-p_1)}.$$

Note that $T^0 = \frac{\|(\Pi_1-\Pi_0)\,\varepsilon\|^2/(p_1-p_0)}{\|(I-\Pi_1)\,\varepsilon\|^2/(n-p_1)}$ whether $\mathbf{H}_0$ is true or not and that if $\beta \in \mathbf{H}_0$ then $T = T^0$. Therefore bootstrap and wild bootstrap estimate the distribution of T consistently on the hypothesis $\mathbf{H}_0$. On the alternative $\mathbf{H}_1$ the wild bootstrap and the bootstrap estimate converges to the distribution of T under a distribution with $\beta = 0 \in \mathbf{H}_0$. For the validity of the approximation by an F distribution in general one has to assume additionally that $E(\varepsilon_i^2|X_i) = E\varepsilon_i^2$ which is not necessary for the bootstrap and the wild bootstrap.

5.Proofs. We will give here only the proof of Theorem 1 which contains the basic argument why under our assumptions an asympitic approach with increasing dimension can be used. The proofs of the other theorems can be found in Mammen ((1989b).

PROOF OF THEOREM 1. The basic idea of the proof is to show that $\widehat{\beta}$ admits the following expansion

$$\sqrt{n}\ \|\widehat{\beta} - \widehat{\beta}_K\| = o_P(1),$$

where $\widehat{\beta}_K$ is the following higher order expansion

$$\widehat{\beta}_K = \beta + \sum_{0 \le h \le K-1} A^h \frac{1}{\sqrt{n}} \sum_{i=1}^{n} X_i \varepsilon_i$$

with $A = I - 1/n \sum_{i=1}^{n} X_i X_i^T$. The integer K is defined in (2.4). The validity of this stochastic expansion follows from

$$\lambda_{amax}(A) = O_P\left(\frac{p^{1/2+\delta/4}}{n^{1/2}}\right) \text{ (see Lemma 1)},$$

where $\lambda_{amax}(A)$ denotes the (random) maximal absolute eigenvalue of A. To see why this bound on $\lambda_{amax}(A)$ implies the stochastic expansion of $\widehat{\beta}$ note first that

$$\frac{1}{\sqrt{n}} \sum_{i=1}^{n} X_i \varepsilon_i = O_P(p^{1/2}),$$

because of $E \| \frac{1}{\sqrt{n}} \sum_{i=1}^{n} X_i \varepsilon_i \|^2 = E \|X_i\|^2 \varepsilon_i^2 = O(p)$ (see (2.4)). This yields

$$\sqrt{n}\ \|\widehat{\beta} - \widehat{\beta}_K\|$$

$$= \sqrt{n}\ \|(1/n \sum_{i=1}^{n} X_i X_i^T)^{-1} n^{-1} \sum_{i=1}^{n} X_i \varepsilon_i - \widehat{\beta}_K\|$$

$$= \|\sum_{h \geq 0} A^h n^{-1/2} \sum_{i=1}^{n} X_i \varepsilon_i - \sqrt{n}\,(\widehat{\beta}_K - \beta)\|$$

$$= \|\sum_{h \geq K} A^h n^{-1/2} \sum_{i=1}^{n} X_i \varepsilon_i \|$$

$$= O_P(\lambda_{amax}(A)^K)\ O_P(p^{1/2})$$

$$= O_P\left(p^{1/2} \left(\frac{p}{n}\right)^{K/2}\right).$$

This is $o_P(1)$ because of (2.3). Below the approximation of $\widehat{\beta}$ by $\widehat{\beta}_K$ will then be used to show

$$\sqrt{n}\ c^T(\widehat{\beta} - E\widehat{\beta}_K) = \frac{1}{\sqrt{n}} \sum_{i=1}^{n} (c^T X_i)\varepsilon_i + o_p(1).$$

Note that the higher order terms of $\widehat{\beta}_K$ appear here only in the expectation.

With analogous arguments we show then for bootstrap and wild bootstrap

$$\sqrt{n}\ c^T(\widehat{\beta}^* - E\widehat{\beta}_K^*) = \frac{1}{\sqrt{n}} \sum_{i=1}^{n} (c^T X_i^*)\varepsilon_i^* + o_p(1),$$

$$\sqrt{n}\ c^T(\widehat{\beta}^W - E\widehat{\beta}_K^W) = \frac{1}{\sqrt{n}} \sum_{i=1}^{n} (c^T X_i)\varepsilon_i^W + o_p(1)$$

(see Lemmas 3* and 6). $\widehat{\beta}_K^*$ and $\widehat{\beta}_K^W$ are higher order stochastic expansions of $\widehat{\beta}^*$ or $\widehat{\beta}^W$ (respectively). $E^*(\ )$ denotes the conditional expectation $E(\ \ |\mathbf{X}, \mathbf{Y})$ and $\varepsilon_i^* = Y_i^* - X_i^{*T} \widehat{\beta}$. By application of the Lindeberg condition (2.5) we will then show that these expressions have the same normal limit. The convergence of the expectations will be ensured by $E(\varepsilon_i|X_i) = 0$ or $\delta \geq 1/3$ ($\delta \geq 1$ respectively). The proof is divided into several lemmas.

LEMMA 0. *Assume* (2.2). *Then*

(5.1) $E(\frac{\|X_i\|}{\sqrt{p}})^{4K}(1+\varepsilon_i^2)$ *is uniformly bounded for* $n \geq 1$.

PROOF. Choose an orthogonal basis $\{e_j : j=1,...,p\}$ of $\mathbf{R}^p$. Then

$$E\,(\,\|X_i\|^2 / p)^{2K}(1+\varepsilon_i^2)$$

$$= E\,(p^{-1}\sum_{j=1}^{p} (e_j^T X_i)^2)^{2K}(1+\varepsilon_i^2)$$

$$= p^{-2K}\sum_{j_1,...,j_{2K}=1}^{p} E\,(e_{j_1}^T X_i)^2 \ldots (e_{j_{2K}}^T X_i)^2\,(1+\varepsilon_i^2)$$

$$\leq \text{const.}\,.$$

LEMMA 1. *Under the assumptions of Theorem* 1

(5.2) $\lambda_{amax}(\,A\,) = O_P(\frac{p^{1/2+\delta/4}}{n^{1/2}})$.

PROOF. We will show

(5.3) $E\Lambda = O(\frac{p^{K+1}}{n^K})$

for $\Lambda = \text{spur}(A^{2K}) \geq \lambda_{amax}(A)^{2K}$. This implies (5.2).

Proof of (5.3). First note that

(5.4) $E\Lambda = n^{-2K} \sum_J \text{ spur } E\,(\,X_{i_1}X_{i_1}^T - I\,)\cdot\ldots\cdot(\,X_{i_{2K}}X_{i_{2K}}^T - I\,)$

where $J = \{(i_1,...,i_{2K}) : \text{For every } j \text{ there exists } h \neq j \text{ with } i_h = i_j\,\}$.

After evaluating the matrix product in (5.4) E Λ is a sum with summands of the following type

(5.5) $n^{-2K}\,S = n^{-2K} \text{ spur } X_{i_1}X_{i_1}^T\cdot\ldots\cdot X_{i_s}X_{i_s}^T = n^{-2K}\,(X_{i_1}^T X_{i_2})\cdot\ldots\cdot(X_{i_s}^T X_{i_1})$

where $0 \leq s \leq 2K$. Put $r = \#\,\{\,i_1,...,i_s\,\}$. To every summand S one can construct a graph with r nodes and s edges. Every node corresponds to an index i_j and two nodes i_h and i_j are connected by an edge if $(X_{i_h}^T X_{i_j})$ appears in S. The order of a node i is the number of edges touching i (where an edge going from i to i (a loop) counts twice). For (r,s) = (K,2K) the order of every node is 4. The graphs are always connected and the order of every node is even.

Now the number of *different* graphs corresponding to the sum in (5.4) is bounded. Furthermore one can show that the number of summands of type (5.5) corresponding to the *same* graph with r nodes and s edges is bounded

by $K^K n^K$ if $s \le 2r$

and by $K^K n^{r+[(2K-s)/2]}$ if $s \ge 2r$.

Here [x] denotes $\sup (n \in \mathbf{N} : n \le x)$. For $s \le 2r$ this follows immediately because the maximal number of different indices in J is K and because for every index j there exist less than K possibilities to choose $h \ne j$ with $i_h = i_j$. For $s \ge 2r$ note that after replacement of *two* factors $\left(X_{i_j}X_{i_j}^T - I\right), \left(X_{i_h}X_{i_h}^T - I\right)$ by I (at most) *one* index may disappear in $\operatorname{spur} (X_{i_1}X_{i_1}^T - I)\bullet\ldots\bullet(X_{i_{2K}}X_{i_{2K}}^T - I)$. Therefore if $s \ge 2r$ the term S corresponds to a summand in (5.4) with an index $(i_1,\ldots,i_{2K}) \in J$ such that $\#\{i_1,\ldots,i_{2K}\} \le r+[(2K-s)/2]$.

We will show

$$(5.6) \qquad E(S) = O(p^{s+1-r}).$$

This implies the statement of Lemma 1 because of :

$$(5.7) \qquad n^{-2K} p^{s+1-r} n^K \le p^{K+1} n^{-K} \qquad \text{if } s \le 2r$$

and

$$n^{-2K} p^{s+1-r} n^{r+[(2K-s)/2]} \le p^{K+1} n^{-K} \left(p^{s-r-K} / n^{s/2-r}\right) \le p^{K+1} n^{-K} \qquad \text{if } s \ge 2r .$$

Proof of (5.6). Define

$$Q = \{i_1,\ldots,i_s\},$$

$$\tilde{Q} = \{i \in Q : \text{There are exactly two edges connecting } i \text{ with another node}\}.$$

First consider the case $\tilde{Q} = \emptyset$. From (2.4) one can follow

(5.8) $$ES = O(p^{\#loops} p^{(1/2)(s-\#loops)}) = O(p^{(1/2)(s+\#loops)}),$$

where #loops denotes the number of loops in the graph. Suppose that there is no element in $\tilde{Q}$. Then for every node i there are at least four edges connecting i with another node (Note that the order of every node is even.). Then there must be at least 2r edges which are no loops. This implies #loops $\leq$ s-2r and $ES = O(p^{s-r}) = O(p^{s-r+1})$.

Suppose now that $\tilde{Q} \neq \emptyset$. Then there exists an $i_0 \in \tilde{Q}$. Denote the neighbors of i_0 by i_1 and i_2 (we do not exclude the case $i_1 = i_2$). Suppose that there are $h \geq 0$ loops are going from i_0 to i_0. Now note that

(5.9) $$E((X_{i_0}^T X_{i_1})(X_{i_0}^T X_{i_2})(X_{i_0}^T X_{i_0})^h \mid X_{i_1}, X_{i_2}) = (X_{i_1}^T M X_{i_2})\, p^h$$

where M is a p×p matrix with bounded maximal absolute eigenvalue (uniformly in i_0 and n). Without loss of generality assume M = I. Then

(5.10) $$E(S \mid X_i : i \neq i_0) = p^h \tilde{S}$$

where $\tilde{S}$ is of type (5.5). The graph which corresponds to $\tilde{S}$ has r-1 nodes and s-h-1 edges. It can be generated from the graph of S by removing the node i_0 and all edges which touch i_0 and by adding an edge which goes from i_1 to i_2. Because of (5.10) it suffices to prove (5.6) for r = 1.

LEMMA 2. *Under the assumptions of Theorem* 1

(5.11) $$\operatorname{var}\left(c^T A^h \frac{1}{\sqrt{n}} \sum_{i=1}^n X_i \varepsilon_i\right) = O\left(\left(\frac{p}{n}\right)^h\right),$$

(5.12) $$E\left(c^T A^h \frac{1}{\sqrt{n}} \sum_{i=1}^n X_i \varepsilon_i\right) = O\left(\frac{p^k}{n^{k-1/2}}\right)$$

for $1 \leq h \leq K-1$ and $k = [\frac{h}{2}]+1$.

PROOF. Note that

$$c^T A^h \frac{1}{\sqrt{n}} \sum_{i=1}^n X_i \varepsilon_i$$

$$= \sum\nolimits^{\neq}_{1\le i_1,\dots,i_q\le n,\ q\le h+1} \sum\nolimits_{k(i_1)+\dots+k(i_q)=h+1} c^T(X_{i_1}X_{i_1}^T - I)^{k(i_1)} \dots (X_{i_q}X_{i_q}^T - I)^{k(i_q)-1} X_{i_q} \varepsilon_{i_q}$$

where $\sum^{\neq}$ denotes summation over pairwise different indices.

The statements of the lemma follow by straightforward but lengthy computations after replacing $\Delta_j = (X_{i_j}X_{i_j}^T - I)^{k(i_j)}$ by $[\Delta_j - E\Delta_j] + E\Delta_j$.

LEMMA 3. *Under the assumptions of Theorem* 1

(5.13) $$\sqrt{n}\, c^T(\hat{\beta} - E\,\hat{\beta}_K) = c^T \frac{1}{\sqrt{n}} \sum_{i=1}^n X_i\varepsilon_i + o_p(1)$$

where $\hat{\beta}_K = \sum_{0\le h\le K-1} A^h \frac{1}{\sqrt{n}} \sum_{i=1}^n X_i\varepsilon_i$.

PROOF. Use $(\frac{1}{n}\sum_{i=1}^n X_iX_i^T)^{-1} = (I - A)^{-1} = I + A + A^2 + \dots$. Because of Lemma 1

(5.14) $$|c^T \sum_{h\ge K} A^h \frac{1}{\sqrt{n}} \sum_{i=1}^n X_i\varepsilon_i|$$

$$\le \lambda_{amax}(A^K) \frac{1}{1-\lambda_{amax}(A)} \left\| \frac{1}{\sqrt{n}} \sum_{i=1}^n X_i\varepsilon_i \right\|$$

$$= O_p\left((p^{1/2+\delta/4}\, n^{-1/2})^K\, p^{1/2}\right)$$

$$= O_p\left((p^{1+\delta}\, n^{-1})^{K/2}\right)$$

$$= o_p(1).$$

Because of Lemma 2

(5.15) $$|c^T \sum_{1\le h\le K-1} A^h \frac{1}{\sqrt{n}} \sum_{i=1}^n X_i\varepsilon_i - EA^h \frac{1}{\sqrt{n}} \sum_{i=1}^n X_i\varepsilon_i| = o_p(1).$$

LEMMA 4. *Under the assumptions of Theorem* 1

$$d_\infty(\mathcal{L}(\sqrt{n}c^T(\hat{\beta}-E\hat{\beta}_K), N(0,E(c^TX_i)^2\varepsilon_i^2) \to 0. \tag{5.15}$$

PROOF. Use condition (2.5) and Lemma 3.

Now we state the bootstrap counterparts of Lemmas 1,..., 4.

LEMMA 1*. *Under the conditions of Theorem* 1

$$\lambda_{amax}(A^*)=O_p(\frac{p^{1/2+\delta/4}}{n^{1/2}}) \tag{5.16}$$

where $A^* = I - \frac{1}{n}\sum_{i=1}^{n} X_i^*(X_i^*)^T$.

PROOF. Similarly as in the proof of Lemma 1 we consider $\Lambda^* = \text{spur}(A^*)^{2K}$. Note that $E\Lambda^*$ is a sum of summands of type (5.5). This sum can be treated as in the proof of Lemma 1.

LEMMA 2*. *Under the assumptions of Theorem* 1

$$\|c^TA^{*h}\frac{1}{\sqrt{n}}\sum_{i=1}^{n} X_i^*\varepsilon_i^* - E^*c^TA^{*h}\frac{1}{\sqrt{n}}\sum_{i=1}^{n} X_i^*\varepsilon_i^*\| = o_p(1) \textit{ for } 0 \le h \le K-1. \tag{5.17}$$

PROOF. See the proof of Lemma 2.

LEMMA 3*. *Under the assumptions of Theorem* 1

$$\sqrt{n}\, c^T(\hat{\beta}^* - E\, \hat{\beta}_K^*) = c^T\frac{1}{\sqrt{n}}\sum_{i=1}^{n} X_i^*\varepsilon_i^* + o_p(1), \tag{5.18}$$

where $\hat{\beta}_K^* = \sum_{0\le h\le K-1} A^{*h}\frac{1}{\sqrt{n}}\sum_{i=1}^{n} X_i^*\varepsilon_i^*$.

LEMMA 4*. *Under the assumptions of Theorem* 1

$$d_\infty(\mathcal{L}^*(\sqrt{n}c^T(\hat{\beta}^* - E\hat{\beta}_K^*), N(0,E(c^TX_i)^2\varepsilon_i^2) \to 0 \textit{ (in probability)}. \tag{5.19}$$

PROOF. To every i there exists a j such that $\varepsilon_i^* = \varepsilon_j - X_j^T(\widehat{\beta} - \beta)$. Put $\varepsilon_{i,1}^* = \varepsilon_j$ and $\varepsilon_{i,2}^* = -X_j^T(\widehat{\beta} - \beta)$. Then $n^{-1/2} \sum_{i=1}^n (c^T X_i^*)\varepsilon_i^* = T_1 + T_2$ where $T_1 = \frac{1}{\sqrt{n}}\sum_{i=1}^n (c^T X_i^*)\varepsilon_{i,1}^*$ and $T_2 = \frac{1}{\sqrt{n}}\sum_{i=1}^n (c^T X_i^*)\varepsilon_{i,2}^*$. First we show

(5.20) $\qquad T_2 - E^*T_2 \rightarrow 0$ (in probability).

Proof of (5.20). First note that

$$\mathrm{var}^*(T_2) = E^*(T_2 - E^*T_2)^2$$

$$\leq \frac{1}{n}\sum_{i=1}^{n} (c^T X_i)^2\, X_i^T(\widehat{\beta}-\beta))^2$$

$$= (\widehat{\beta}-\beta)^T \frac{1}{n}\sum_{i=1}^{n} Z_i Z_i^T\, (\widehat{\beta}-\beta)$$

for $Z_i = (c^T X_i)X_i$. Using $\|\widehat{\beta}-\beta\|^2 = O_p(p/n)$ and using that $E(\|Z_i\|/\sqrt{p})^{2K}$ and $E(d^T Z_i)^{2K}$ are uniformly bounded for $\|d\| = 1$ and $n \geq 1$ (see (2.2) and Lemma 0) one can show with the same arguments as in the proof of Lemma 1 that $\lambda_{amax}(\frac{1}{n}\sum_{i=1}^n Z_i Z_i^T - E\, Z_i Z_i^T) = o_p(1)$. This implies $\mathrm{var}^*(T_2) = O_p(p/n)$ and (5.20).

Now note that $\sum_{i=1}^n (c^T X_i)(Y_i - X_i^T\widehat{\beta}) = 0$ implies that

$$E^*T_2 = -\frac{1}{\sqrt{n}}\sum_{i=1}^n (c^T X_i)(X_i^T(\widehat{\beta}-\beta)) = -\frac{1}{\sqrt{n}}\sum_{i=1}^n (c^T X_i)\varepsilon_i = -E^*T_1.$$

Therefore it remains to show the asymptotic normality of $T_1 - E^*T_1$. This can be shown by proving the Lindeberg condition:

$$\exists \gamma_n \rightarrow 0 \text{ such that } \frac{1}{n}\sum_{i=1}^n [(c^T X_i)\varepsilon_i - \Delta]^2 I([(c^T X_i)\varepsilon_i - \Delta]^2 > \gamma_n\, n) \rightarrow 0 \text{ (in probability)}$$

where $\Delta = \frac{1}{n}\sum_{i=1}^n (c^T X_i)\varepsilon_i$. For the proof of the Lindeberg condition first note that $\Delta = o_p(1)$ and that $\frac{1}{n}\sum_{i=1}^n (c^T X_i)^2\varepsilon_i^2\, I((c^T X_i)^2\varepsilon_i^2 > \gamma_n\, n) \rightarrow 0$ (in probability) if γ_n goes slowly enough to 0 (because of (2.7)). This implies for γ_n converging slowly enough to 0

$$\frac{1}{n}\sum_{i=1}^n [(c^T X_i)\varepsilon_i - \Delta]^2 I([(c^T X_i)\varepsilon_i - \Delta]^2 > \gamma_n\, n)$$

$$\leq \frac{1}{n}\sum_{i=1}^{n} [(c^TX_i)\varepsilon_i]^2 I([(c^TX_i)\varepsilon_i-\Delta]^2 > \gamma_n n) + o_P(1)$$

$$\leq \frac{1}{n}\sum_{i=1}^{n} [(c^TX_i)\varepsilon_i]^2$$

$$\left\{ I[\,|(c^TX_i)\varepsilon_i| \geq .5\,\gamma_n^{1/2} n^{1/2}] + I[\,|(c^TX_i)\varepsilon_i| \leq .5\,\gamma_n^{1/2} n^{1/2}, |\Delta| \geq .5\,\gamma_n^{1/2} n^{1/2}]\right\} + o_P(1)$$

$$\leq \frac{1}{n}\sum_{i=1}^{n} .25\,\gamma_n\, n\, I[\,|\Delta| \geq .5\,\gamma_n^{1/2} n^{1/2}] + o_P(1)$$

$$\leq \Delta^2 + o_P(1) = o_P(1).$$

For the proof of the first statement of Theorem 1 it remains to show the following lemma.

LEMMA 5. *Under the assumptions of Theorem* 1

$$E^*(c^T(\hat{\beta}_K^* - \hat{\beta})) - E(c^T(\hat{\beta}_K - \hat{\beta})) = o_p(1/\sqrt{n}) \tag{5.21}$$

if $\delta \geq 1/3$ *or* $E(\varepsilon_i|X_i) = 0$.

PROOF. Assume first $E(\varepsilon_i|X_i) = 0$. Then $E(c^T\hat{\beta}_K) = c^T\beta$. Furthermore by straightforward calculations one can show $E^*(c^T\hat{\beta}_K^*) = c^T\hat{\beta} + o_p(1)$. This shows (5.21).

Assume now $\delta \geq 1/3$. Then $p^2 n^{-3/2} = o(1)$ and (see Lemma 2)

$$\sqrt{n}E(c^T\hat{\beta}_K - \beta) = E\, c^T(I+A)\frac{1}{\sqrt{n}}\sum_{i=1}^{n} X_i\varepsilon_i + O(p^2 n^{-3/2})$$

$$= \frac{1}{\sqrt{n}}E(c^TX_1)\,\|X_1\|^2\,\varepsilon_1 + o(1).$$

By similar calculations one gets

$$\sqrt{n}E^*(c^T\hat{\beta}_K^* - c^T\hat{\beta}) = \sqrt{n}\,\frac{1}{n}\sum_i (c^TX_i)\|X_i\|^2\varepsilon_i + O_P(p^2 n^{-3/2})$$

$$= \sqrt{n}E(c^T\hat{\beta}_K - c^T\beta) + o_P(1).$$

But this implies (5.21).

We will now prove the second statement of Theorem 1 - the consistency of wild bootstrap.

LEMMA 6. *Under the conditions of Theorem* 1

$$\sqrt{n}c^T(\hat{\beta}^W - \hat{\beta}) = \frac{1}{\sqrt{n}}\sum_{i=1}^{n} (c^T X_i)\varepsilon_i^W + o_p(1) \tag{5.22}$$

if $E(\varepsilon_i|X_i) = 0$ *or* $\delta \geq 1$.

PROOF. The proof goes along the lines of the proofs of Lemmas 1, 2, and 3. Note that $\sqrt{n}E^*c^T(\hat{\beta}^W - \hat{\beta}) = o_P(1)$ (see Lemma 2).

LEMMA 6*. *Under the conditions of Theorem* 1

$$d_\infty(\mathcal{L}^*(\sqrt{n}c^T(\hat{\beta}^W - \hat{\beta})), N(0, E(c^T X_i)^2\varepsilon_i^2)) \to 0 \textit{ (in probability)} \tag{5.23}$$

if $E(\varepsilon_i \mid X_i) = 0$ *or* $\delta \geq 1$.

PROOF. In the first part of the proof we will show that

$$\text{var}^*(\sqrt{n}c^T(\hat{\beta}^W - \hat{\beta})) \to E(c^T X_i)^2\varepsilon_i^2. \tag{5.24}$$

In the second part we will prove the asymptotic normality of $\mathcal{L}^*(\sqrt{n}c^T(\hat{\beta}^W - \hat{\beta}))$. First note that

$$\text{var}^*(\sqrt{n}c^T(\hat{\beta}^W - \hat{\beta})) = \frac{1}{n}\sum_{i=1}^n E^*(c^T X_i)^2(\varepsilon_i^W)^2$$

$$= \frac{1}{n}\sum_{i=1}^n (c^T X_i)^2(\varepsilon_i - X_i^T(\hat{\beta} - \beta))^2 .$$

Now

$$\frac{1}{n}\sum_{i=1}^n \left(c^T X_i\right)^2 \left(X_i^T(\hat{\beta} - \beta)\right)^2 = o_p(1) \text{ (see the proof of Lemma 4*)}. \tag{5.25}$$

Furthermore we will show

(5.26) $$\frac{1}{n}\sum_{i=1}^{n} (c^TX_i)^2\varepsilon_i^2 \to E(c^TX_i)^2\varepsilon_i^2.$$

But (5.25) and (5.26) imply (5.24).

Proof of (5.26). First choose $\gamma_n \to 0$ such that

$$E(c^TX_i)^2\varepsilon_i^2 \; I((c^TX_i)^2\varepsilon_i^2 \geq \gamma_n\, n) \to 0.$$

Choose $c_n \to 0$, $c_n > \gamma_n$ such that

$$nP((c^TX_i)^2\varepsilon_i^2 > c_n n) \leq E(c^TX_i)^2\varepsilon_i^2/c_n \; I((c^TX_i)^2\varepsilon_i^2 \geq \gamma_n\, n) \to 0.$$

Then $\sup_{1\leq i\leq n} (c^TX_i)^2\varepsilon_i^2 = o_p(c_n n)$ and $\frac{1}{n}\sum_{i=1}^{n} (c^TX_i)^2\varepsilon_i^2 - ((c^TX_i)^2\varepsilon_i^2 \wedge c_n n) \to 0$ (in probability) ($a\wedge b$ denotes the infimum of a and b).

Now

$$\operatorname{var}\left(\frac{1}{n}\sum_{i=1}^{n} ((c^TX_i)^2\varepsilon_i^2 \wedge c_n n)\right) \leq \frac{1}{n}E(c^TX_i)^2\varepsilon_i^2 c_n n \to 0.$$

This implies (5.26).

For the proof of asymptotic normality of $\mathcal{L}^*(\sqrt{n}c^T(\hat{\beta}^W - \hat{\beta}))$ it suffices to show for $\gamma > 0$:

$$\frac{1}{n}\sum_{i=1}^{n} E^*(c^TX_i)^2(\varepsilon_i^W)^2 \; I((c^TX_i)^2(\varepsilon_i^W)^2 \geq \gamma\, n) \to 0 \text{ (in probability)}.$$

But this follows similarly as in the proof of Lemma 4* from

$$\frac{1}{n}\sum_{i=1}^{n} (c^TX_i)^2(\varepsilon_i)^2 \; I((c^TX_i)^2\varepsilon_i^2 \geq \gamma\, n) = o_p(1), \text{ for } \gamma > 0$$

because of

$$|\varepsilon_i^W|^2 \leq \text{const. } \hat{\varepsilon}_i^2 \text{ or } |\varepsilon_i^W|^2 \leq \text{const. } U_i^2\, \hat{\varepsilon}_i^2$$

where $U_1,\ldots,U_n$ are i.i.d. variables with distribution as described in the introduction.

	10 % quantile		90 % quantile	
Bootstrap	0.093	0.100	0.895	0.893
	0.031	0.018	0.033	0.017
Wild	0.080	0.096	0.913	0.889
bootstrap	0.027	0.010	0.028	0.011
Normal	0.094	0.076	0.900	0.902
approximation	0.028		0.029	

Table 2a. Monte Carlo estimates of the expectation (first row) and standard deviation (second row) of estimates of the distribution function of an unstudentized linear contrast (columns 1 and 3) and a studentized linear contrast (column 2 and 4) at the 10% and 90% quantiles. (Model A, $p = 5$, $c = c_1$)

	10 % quantile		90 % quantile	
Bootstrap	0.106	0.096	0.893	0.896
	0.050	0.024	0.050	0.024
Wild	0.082	0.091	0.917	0.900
bootstrap	0.043	0.015	0.043	0.015
Normal	0.100	0.070	0.900	0.921
approximation	0.050		0.050	

Table 2b. Monte Carlo estimates of the expectation (first row) and standard deviation (second row) of estimates of the distribution function of an unstudentized linear contrast (columns 1 and 3) and a studentized linear contrast (column 2 and 4) at the 10% and 90% quantiles. (Model A, $p = 5$, $c = c_2$)

	10 % quantile		90 % quantile	
Bootstrap	0.109	0.096	0.907	0.910
	0.037	0.024	0.036	0.021
Wild	0.073	0.082	0.941	0.913
bootstrap	0.028	0.011	0.025	0.011
Normal	0.096	0.055	0.923	0.936
approximation	0.032		0.029	

Table 2c. Monte Carlo estimates of the expectation (first row) and standard deviation (second row) of estimates of the distribution function of an unstudentized linear contrast (columns 1 and 3) and a studentized linear contrast (column 2 and 4) at the 10% and 90% quantiles. (Model A, $p = 10$, $c = c_1$)

	10 % quantile		90 % quantile	
Bootstrap	0.111	0.094	0.878	0.890
	0.052	0.028	0.050	0.027
Wild	0.062	0.080	0.927	0.900
bootstrap	0.038	0.016	0.040	0.016
Normal	0.084	0.052	0.902	0.931
	0.046		0.048	

Table 2d. Monte Carlo estimates of the expectation (first row) and standard deviation (second row) of estimates of the distribution function of an unstudentized linear contrast (columns 1 and 3) and a studentized linear contrast (column 2 and 4) at the 10% and 90% quantiles. (Model A, $p = 10$, $c = c_2$)

	10 % quantile		90 % quantile	
Bootstrap	0.083	0.094	0.900	0.866
	0.073	0.063	0.071	0.062
Wild	0.068	0.075	0.919	0.877
bootstrap	0.053	0.014	0.057	0.017
Normal	0.084	0.057	0.899	0.903
approximation	0.063		0.067	

Table 3a. Monte Carlo estimates of the expectation (first row) and standard deviation (second row) of estimates of the distribution function of an unstudentized linear contrast (columns 1 and 3) and a studentized linear contrast (column 2 and 4) at the 10% and 90% quantiles. (Model B, $p = 5$, $c = c_1$)

	10 % quantile		90 % quantile	
Bootstrap	0.080	0.112	0.906	0.870
	0.085	0.076	0.089	0.079
Wild	0.060	0.080	0.928	0.896
bootstrap	0.061	0.024	0.067	0.027
Normal	0.084	0.044	0.900	0.940
approximation	0.080		0.085	

Table 3b. Monte Carlo estimates of the expectation (first row) and standard deviation (second row) of estimates of the distribution function of an unstudentized linear contrast (columns 1 and 3) and a studentized linear contrast (column 2 and 4) at the 10% and 90% quantiles. (Model B, $p = 5$, $c = c_2$)

	10 % quantile		90 % quantile	
Bootstrap	0.107	0.110	0.916	0.873
	0.075	0.061	0.056	0.055
Wild	0.069	0.090	0.949	0.891
bootstrap	0.046	0.017	0.040	0.017
Normal	0.092	0.059	0.929	0.927
approximation	0.057		0.051	

Table 3c. Monte Carlo estimates of the expectation (first row) and standard deviation (second row) of estimates of the distribution function of an unstudentized linear contrast (columns 1 and 3) and a studentized linear contrast (column 2 and 4) at the 10% and 90% quantiles. (Model B, $p = 10, c = c_1$)

	10 % quantile		90 % quantile	
Bootstrap	0.084	0.108	0.919	0.898
	0.075	0.068	0.069	0.061
Wild	0.044	0.077	0.956	0.927
bootstrap	0.044	0.023	0.044	0.023
Normal	0.070	0.035	0.931	0.966
approximation	0.061		0.061	

Table 3d. Monte Carlo estimates of the expectation (first row) and standard deviation (second row) of estimates of the distribution function of an unstudentized linear contrast (columns 1 and 3) and a studentized linear contrast (column 2 and 4) at the 10% and 90% quantiles. (Model B, $p = 10, c = c_2$)

	10 % quantile		90 % quantile	
Bootstrap	0.089	0.076	0.900	0.876
	0.070	0.041	0.036	0.019
Wild	0.028	0.009	0.848	0.769
bootstrap	0.028	0.003	0.050	0.016
Normal	0.046	0.005	0.821	0.791
approximation	0.039		0.058	

Table 4a. Monte Carlo estimates of the expectation (first row) and standard deviation (second row) of estimates of the distribution function of an unstudentized linear contrast (columns 1 and 3) and a studentized linear contrast (column 2 and 4) at the 10% and 90% quantiles. (Model C, $p = 5, c = c_1$)

	10 % quantile		90 % quantile	
Bootstrap	0.069	0.102	0.928	0.899
	0.068	0.064	0.068	0.065
Wild	0.049	0.077	0.949	0.925
bootstrap	0.044	0.023	0.045	0.023
Normal	0.073	0.040	0.924	0.962
approximation	0.063		0.064	

Table 4b. Monte Carlo estimates of the expectation (first row) and standard deviation (second row) of estimates of the distribution function of an unstudentized linear contrast (columns 1 and 3) and a studentized linear contrast (column 2 and 4) at the 10% and 90% quantiles. (Model C, $p = 5, c = c_2$)

	10 % quantile		90 % quantile	
Bootstrap	0.065	0.064	0.867	0.849
	0.055	0.034	0.030	0.023
Wild	0.007	0.003	0.777	0.698
bootstrap	0.011	0.002	0.043	0.016
Normal	0.017	0.001	0.741	0.729
approximation	0.020		0.045	

Table 4c. Monte Carlo estimates of the expectation (first row) and standard deviation (second row) of estimates of the distribution function of an unstudentized linear contrast (columns 1 and 3) and a studentized linear contrast (column 2 and 4) at the 10% and 90% quantiles. (Model C, $p = 10$, $c = c_1$)

	10 % quantile		90 % quantile	
Bootstrap	0.071	0.083	0.905	0.895
	0.055	0.050	0.061	0.053
Wild	0.032	0.055	0.951	0.920
bootstrap	0.029	0.019	0.037	0.023
Normal	0.054	0.022	0.922	0.963
approximation	0.044		0.052	

Table 4d. Monte Carlo estimates of the expectation (first row) and standard deviation (second row) of estimates of the distribution function of an unstudentized linear contrast (columns 1 and 3) and a studentized linear contrast (column 2 and 4) at the 10% and 90% quantiles. (Model C, $p = 10$, $c = c_2$)

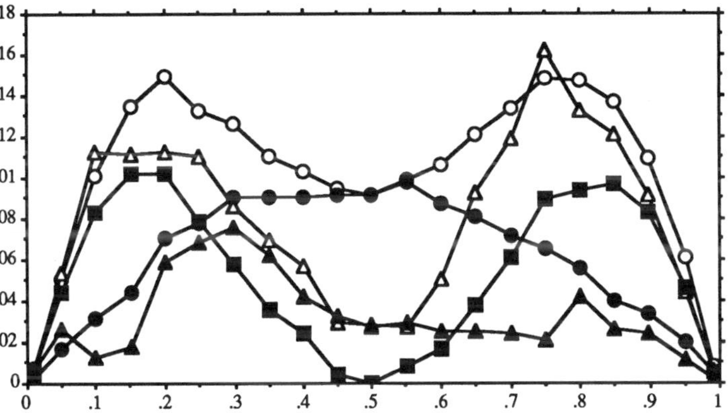

Figure 1aa. *Monte Carlo estimates of the* ***expected squared error*** *of estimates of the distribution function of a unstudentized linear contrast and a studentized linear contrast, plotted against the distribution function. (Model A,* p = 5, c = c1 *).* ○ *bootstrap,* △ *wild bootstrap,* ■ *normal approximation of the unstudentized linear contrast,* ● *bootstrap of the studentized statistic,* ▲ *wild bootstrap of the studentized statistic.*

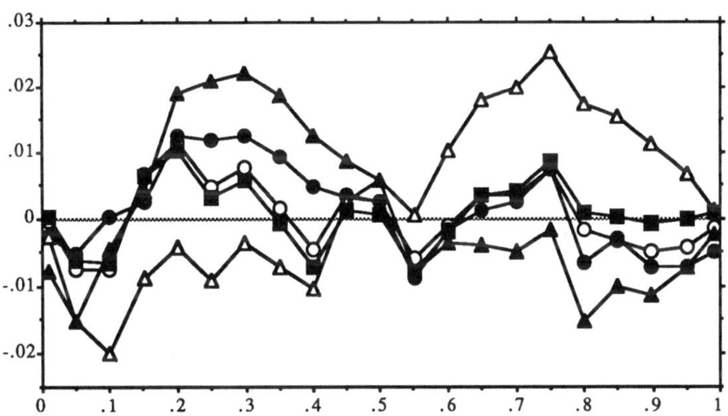

Figure 1ab. *Monte Carlo estimates of the* ***bias*** *of estimates of the distribution function of a unstudentized linear contrast and a studentized linear contrast, plotted against the distribution function. M estimation with scale estiomation. (Model A,* p = 5, c = c1 *).* ○ *bootstrap,* △ *wild bootstrap,* ■ *normal approximation of the unstudentized linear contrast,* ● *bootstrap of the studentized statistic,* ▲ *wild bootstrap of the studentized statistic.*

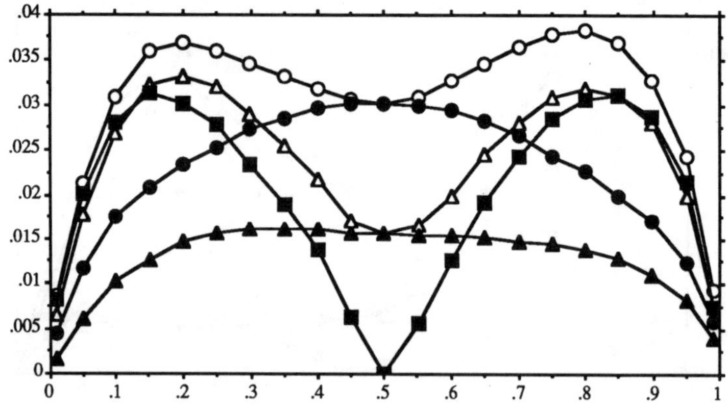

Figure 1ac. *Monte Carlo estimates of the* **standard deviation** *of estimates of the distribution function of a unstudentized linear contrast and a studentized linear contrast , plotted against the distribution function. (Model A,* $p = 5$, $c = c_1$ *).* ○ *bootstrap,* △ *wild bootstrap,* ■ *normal approximation of the unstudentized linear contrast,* ● *bootstrap of the studentized statistic,* ▲ *wild bootstrap of the studentized statistic.*

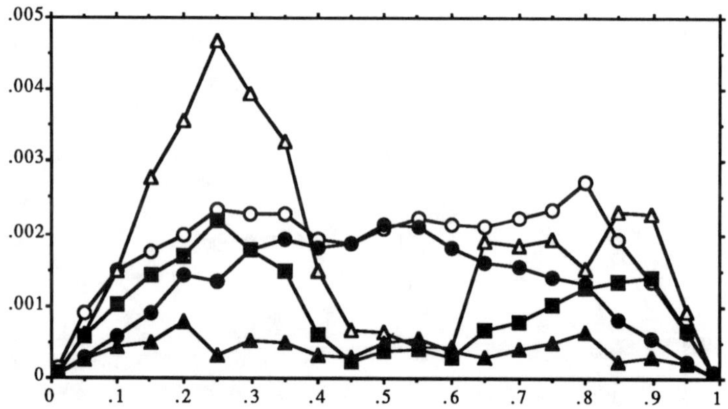

Figure 1ad. *Monte Carlo estimates of the* **expected squared error** *of estimates of the distribution function of a unstudentized linear contrast and a studentized linear contrast , plotted against the distribution function. (Model A,* $p = 10$, $c = c_1$ *).* ○ *bootstrap,* △ *wild bootstrap,* ■ *normal approximation of the unstudentized linear contrast,* ● *bootstrap of the studentized statistic,* ▲ *wild bootstrap of the studentized statistic.*

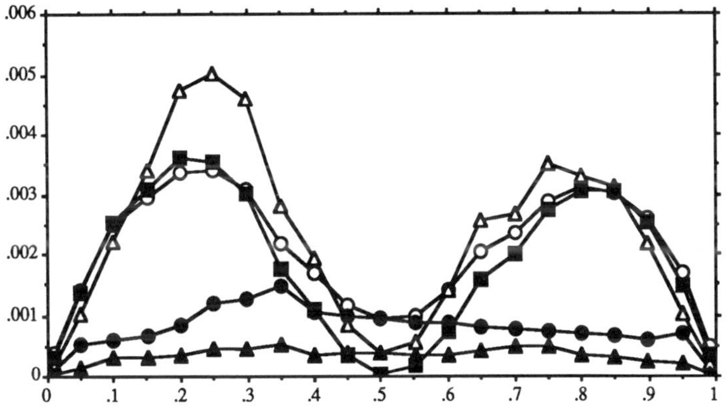

Figure 1ba. *Monte Carlo estimates of the* ***expected squared error*** *of estimates of the distribution function of a unstudentized linear contrast and a studentized linear contrast , plotted against the distribution function. (Model A,* p = 5, c = c_2 *). ○ bootstrap, △ wild bootstrap, ■ normal approximation of the unstudentized linear contrast, ● bootstrap of the studentized statistic, ▲ wild bootstrap of the studentized statistic.*

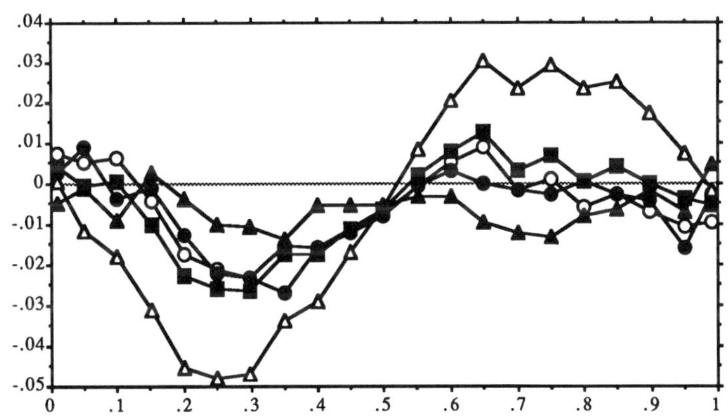

Figure 1bb. *Monte Carlo estimates of the* ***bias*** *of estimates of the distribution function of a unstudentized linear contrast and a studentized linear contrast , plotted against the distribution function. (Model A,* p = 5, c = c_2 *). ○ bootstrap, △ wild bootstrap, ■ normal approximation of the unstudentized linear contrast, ● bootstrap of the studentized statistic, ▲ wild bootstrap of the studentized statistic.*

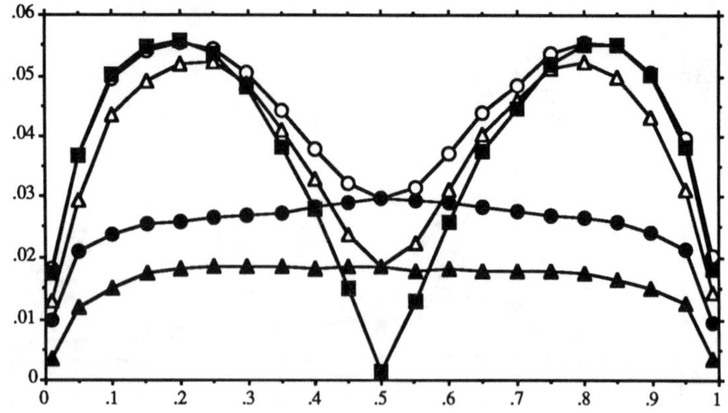

Figure 1bc. *Monte Carlo estimates of the* **standard deviation** *of estimates of the distribution function of a unstudentized linear contrast and a studentized linear contrast , plotted against the distribution function. (Model A,* p = 5, c = c_2 *).* ○ *bootstrap,* △ *wild bootstrap,* ■ *normal approximation of the unstudentized linear contrast,* ● *bootstrap of the studentized statistic,* ▲ *wild bootstrap of the studentized statistic.*

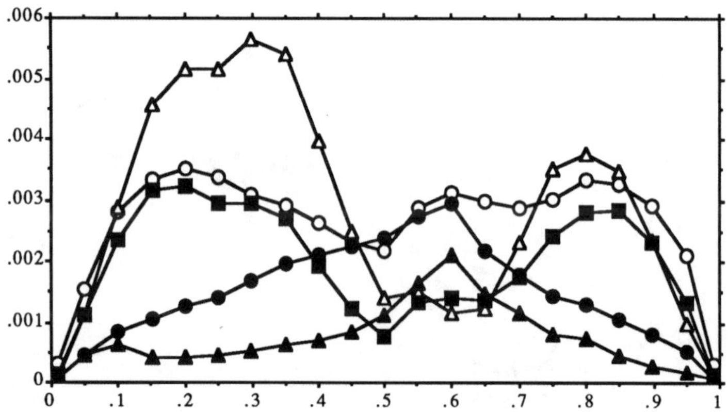

Figure 1bd. *Monte Carlo estimates of the* **expected squared error** *of estimates of the distribution function of a unstudentized linear contrast and a studentized linear contrast , plotted against the distribution function. (Model A,* p = 10, c = c_2 *).* ○ *bootstrap,* △ *wild bootstrap,* ■ *normal approximation of the unstudentized linear contrast,* ● *bootstrap of the studentized statistic,* ▲ *wild bootstrap of the studentized statistic.*

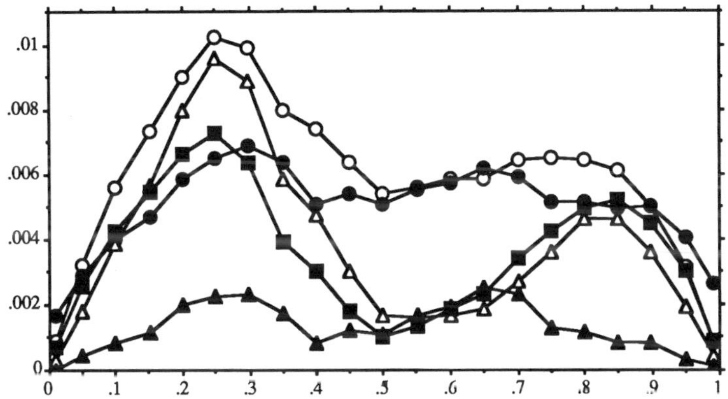

Figure 2aa. *Monte Carlo estimates of the* ***expected squared error*** *of estimates of the distribution function of a unstudentized linear contrast and a studentized linear contrast, plotted against the distribution function. (Model B,* $p = 5$, $c = c_1$ *).* ○ *bootstrap,* △ *wild bootstrap,* ■ *normal approximation of the unstudentized linear contrast,* ● *bootstrap of the studentized statistic,* ▲ *wild bootstrap of the studentized statistic.*

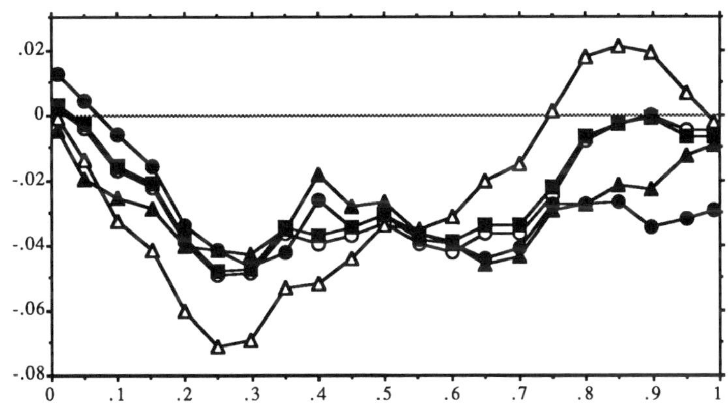

Figure 2ab. *Monte Carlo estimates of the* ***bias*** *of estimates of the distribution function of a unstudentized linear contrast and a studentized linear contrast, plotted against the distribution function. (Model B,* $p = 5$, $c = c_1$ *).* ○ *bootstrap,* △ *wild bootstrap,* ■ *normal approximation of the unstudentized linear contrast,* ● *bootstrap of the studentized statistic,* ▲ *wild bootstrap of the studentized statistic.*

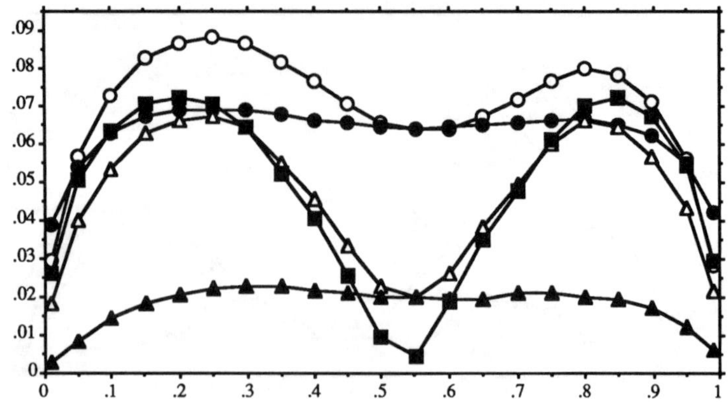

Figure 2ac. *Monte Carlo estimates of the* ***standard deviation*** *of estimates of the distribution function of a unstudentized linear contrast and a studentized linear contrast , plotted against the distribution function.* (*Model B*, p = 5, c = c_1). ○ *bootstrap*, △ *wild bootstrap*, ■ *normal approximation of the unstudentized linear contrast*, ● *bootstrap of the studentized statistic*, ▲ *wild bootstrap of the studentized statistic.*

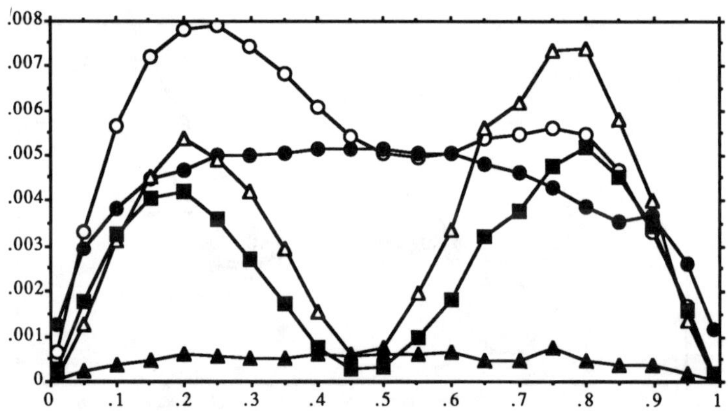

Figure 2ad. *Monte Carlo estimates of the* ***expected squared error*** *of estimates of the distribution function of a unstudentized linear contrast and a studentized linear contrast , plotted against the distribution function.* (*Model B*, p = 10, c = c_1). ○ *bootstrap*, △ *wild bootstrap*, ■ *normal approximation of the unstudentized linear contrast*, ● *bootstrap of the studentized statistic*, ▲ *wild bootstrap of the studentized statistic.*

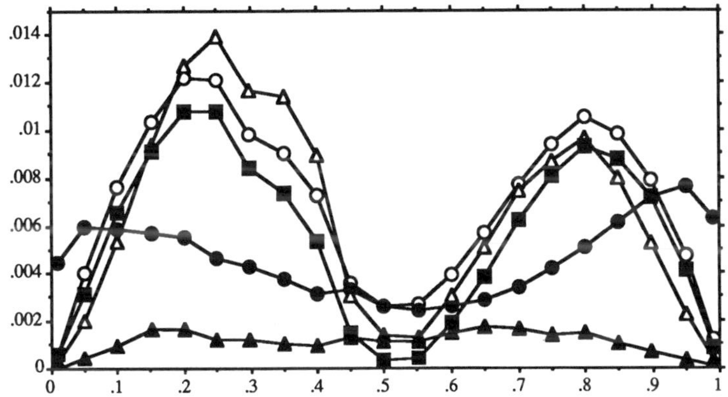

Figure 2ba. *Monte Carlo estimates of the* ***expected squared error*** *of estimates of the distribution function of a unstudentized linear contrast and a studentized linear contrast , plotted against the distribution function. (Model B,* p = 5, c = c_2 *).* ○ *bootstrap,* △ *wild bootstrap,* ■ *normal approximation of the unstudentized linear contrast,* ● *bootstrap of the studentized statistic,* ▲ *wild bootstrap of the studentized statistic.*

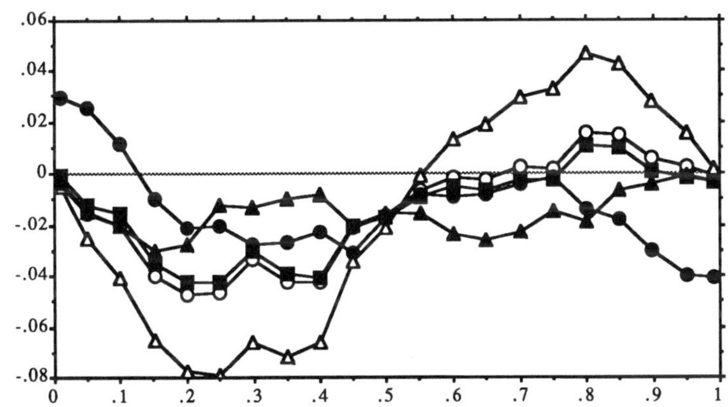

Figure 2bb. *Monte Carlo estimates of the* ***bias*** *of estimates of the distribution function of a unstudentized linear contrast and a studentized linear contrast , plotted against the distribution function. (Model B,* p = 5, c = c_2 *).* ○ *bootstrap,* △ *wild bootstrap,* ■ *normal approximation of the unstudentized linear contrast,* ● *bootstrap of the studentized statistic,* ▲ *wild bootstrap of the studentized statistic.*

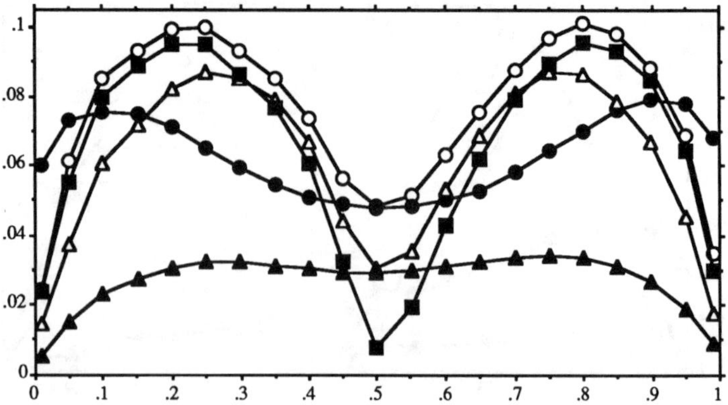

Figure 2bc. *Monte Carlo estimates of the* ***standard deviation*** *of estimates of the distribution function of a unstudentized linear contrast and a studentized linear contrast , plotted against the distribution function. (Model B,* $p = 5, c = c_2$ *).* ○ *bootstrap,* △ *wild bootstrap,* ■ *normal approximation of the unstudentized linear contrast,* ● *bootstrap of the studentized statistic,* ▲ *wild bootstrap of the studentized statistic.*

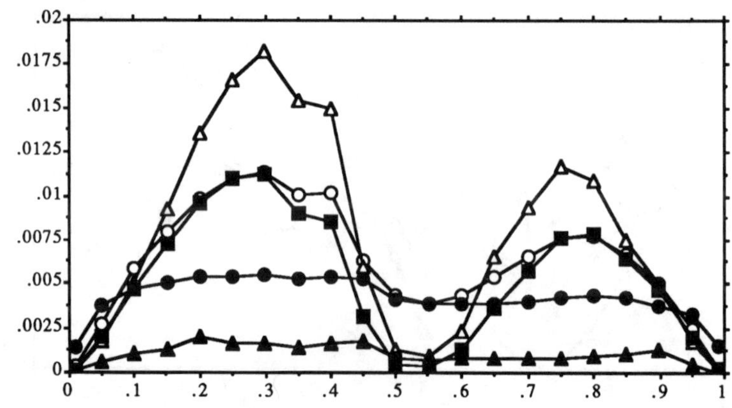

Figure 2bd. *Monte Carlo estimates of the* ***expected squared error*** *of estimates of the distribution function of a unstudentized linear contrast and a studentized linear contrast , plotted against the distribution function. (Model B,* $p = 10, c = c_2$ *).* ○ *bootstrap,* △ *wild bootstrap,* ■ *normal approximation of the unstudentized linear contrast,* ● *bootstrap of the studentized statistic,* ▲ *wild bootstrap of the studentized statistic.*

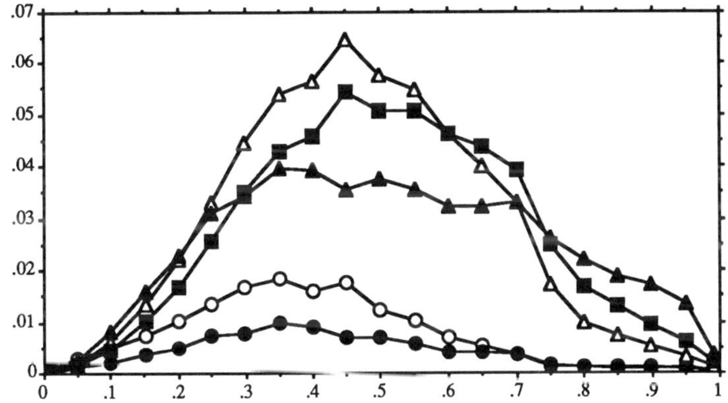

Figure 3aa. *Monte Carlo estimates of the* **expected squared error** *of estimates of the distribution function of a unstudentized linear contrast and a studentized linear contrast, plotted against the distribution function. (Model C,* p = 5, c = c_1 *).* ○ *bootstrap,* △ *wild bootstrap,* ■ *normal approximation of the unstudentized linear contrast,* ● *bootstrap of the studentized statistic,* ▲ *wild bootstrap of the studentized statistic.*

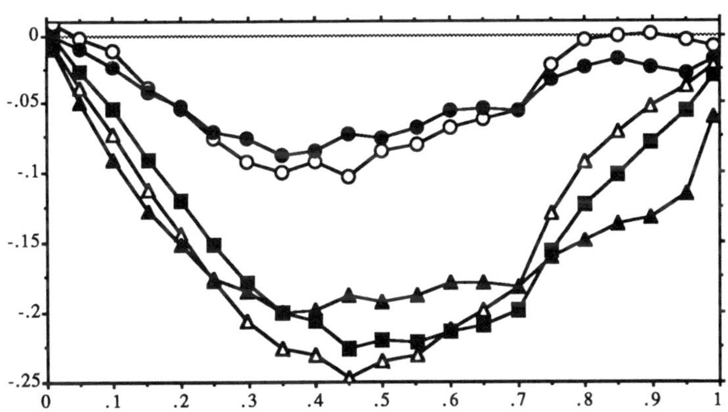

Figure 3ab. *Monte Carlo estimates of the* **bias** *of estimates of the distribution function of a unstudentized linear contrast and a studentized linear contrast, plotted against the distribution function. (Model C,* p = 5, c = c_1 *).* ○ *bootstrap,* △ *wild bootstrap,* ■ *normal approximation of the unstudentized linear contrast,* ● *bootstrap of the studentized statistic,* ▲ *wild bootstrap of the studentized statistic.*

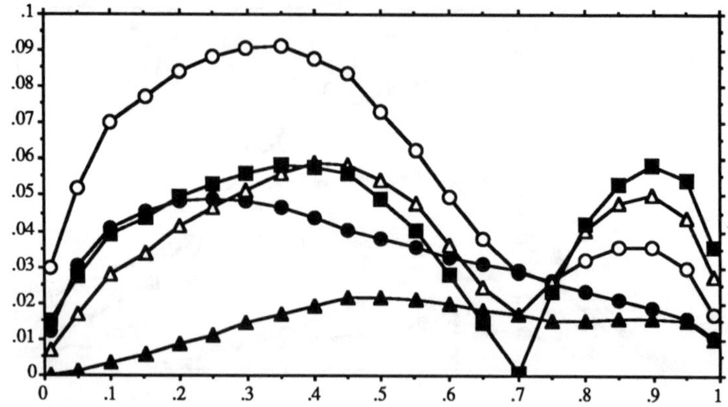

Figure 3ac. *Monte Carlo estimates of the* **standard deviation** *of estimates of the distribution function of a unstudentized linear contrast and a studentized linear contrast , plotted against the distribution function. (Model C,* p = 5, c = c_1 *).* ○ *bootstrap,* △ *wild bootstrap,* ■ *normal approximation of the unstudentized linear contrast,* ● *bootstrap of the studentized statistic,* ▲ *wild bootstrap of the studentized statistic.*

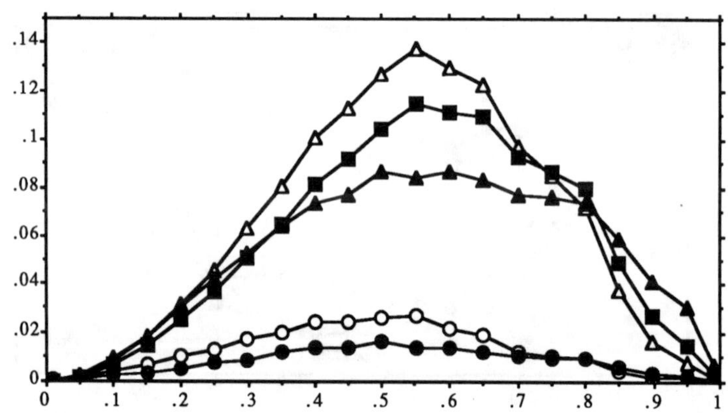

Figure 3ad. *Monte Carlo estimates of the* **expected squared error** *of estimates of the distribution function of a unstudentized linear contrast and a studentized linear contrast , plotted against the distribution function. (Model C,* p = 10, c = c_1 *).* ○ *bootstrap,* △ *wild bootstrap,* ■ *normal approximation of the unstudentized linear contrast,* ● *bootstrap of the studentized statistic,* ▲ *wild bootstrap of the studentized statistic.*

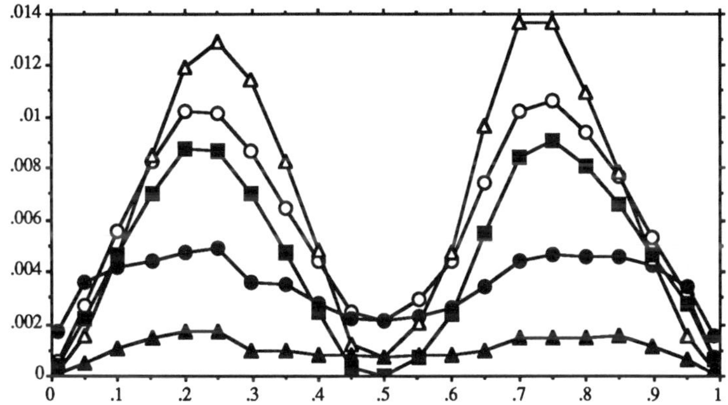

Figure 3ba. *Monte Carlo estimates of the* ***expected squared error*** *of estimates of the distribution function of a unstudentized linear contrast and a studentized linear contrast, plotted against the distribution function. (Model C,* p = 5, c = c_2 *).* ○ *bootstrap,* △ *wild bootstrap,* ■ *normal approximation of the unstudentized linear contrast,* ● *bootstrap of the studentized statistic,* ▲ *wild bootstrap of the studentized statistic.*

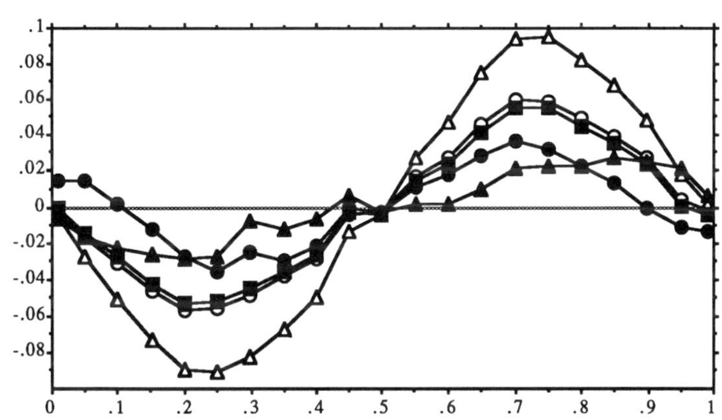

Figure 3bb. *Monte Carlo estimates of the* ***bias*** *of estimates of the distribution function of a unstudentized linear contrast and a studentized linear contrast, plotted against the distribution function. (Model C,* p = 5, c = c_2 *).* ○ *bootstrap,* △ *wild bootstrap,* ■ *normal approximation of the unstudentized linear contrast,* ● *bootstrap of the studentized statistic,* ▲ *wild bootstrap of the studentized statistic.*

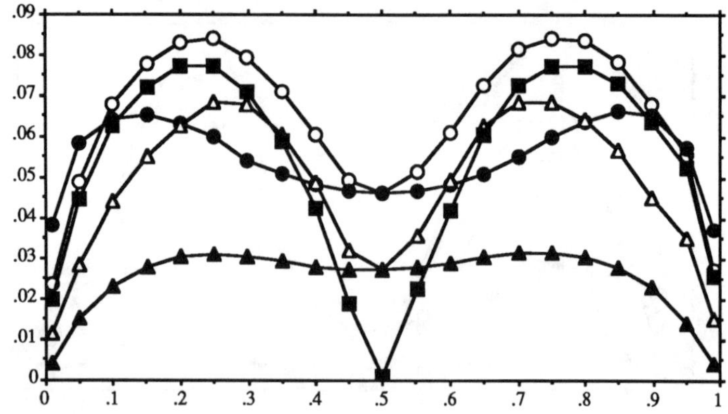

Figure 3bc. *Monte Carlo estimates of the* **standard deviation** *of estimates of the distribution function of a unstudentized linear contrast and a studentized linear contrast , plotted against the distribution function. (Model C,* p = 5, c = c_2 *).* ○ *bootstrap,* △ *wild bootstrap,* ■ *normal approximation of the unstudentized linear contrast,* ● *bootstrap of the studentized statistic,* ▲ *wild bootstrap of the studentized statistic.*

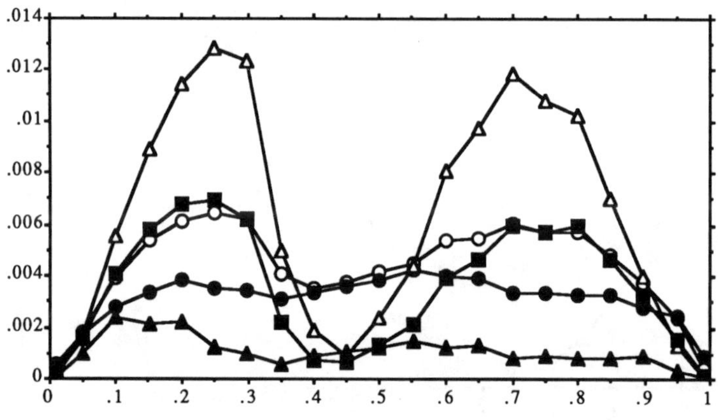

Figure 3bd. *Monte Carlo estimates of the* **expected squared error** *of estimates of the distribution function of a unstudentized linear contrast and a studentized linear contrast , plotted against the distribution function. (Model C,* p = 10, c = c_2 *).* ○ *bootstrap,* △ *wild bootstrap,* ■ *normal approximation of the unstudentized linear contrast,* ● *bootstrap of the studentized statistic,* ▲ *wild bootstrap of the studentized statistic.*

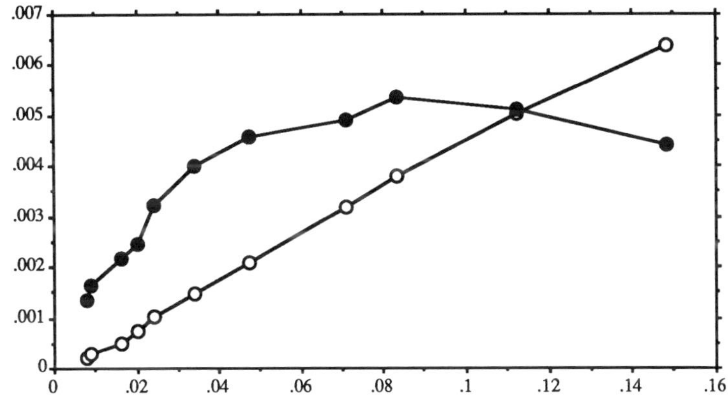

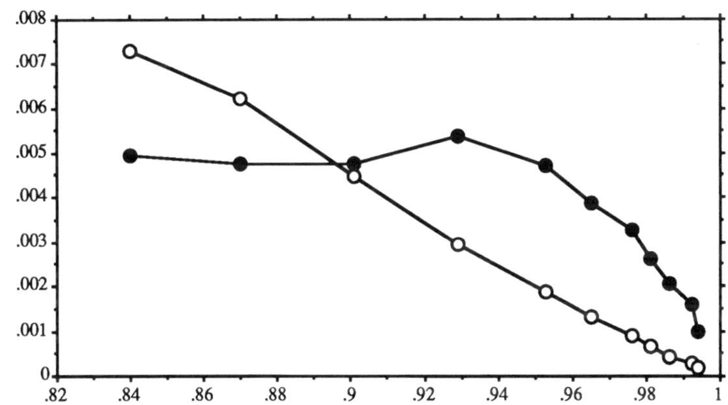

Figure 4a. *Monte Carlo estimates of the* ***expected squared error*** *of estimates of the distribution function of a unstudentized linear contrast and a studentized linear contrast , plotted against the distribution function. (Model D,* $p = 5$, $c = c_2$ *).* ○ *bootstrap of the unstudentized linear contrast,* ● *bootstrap of the studentized statistic.*

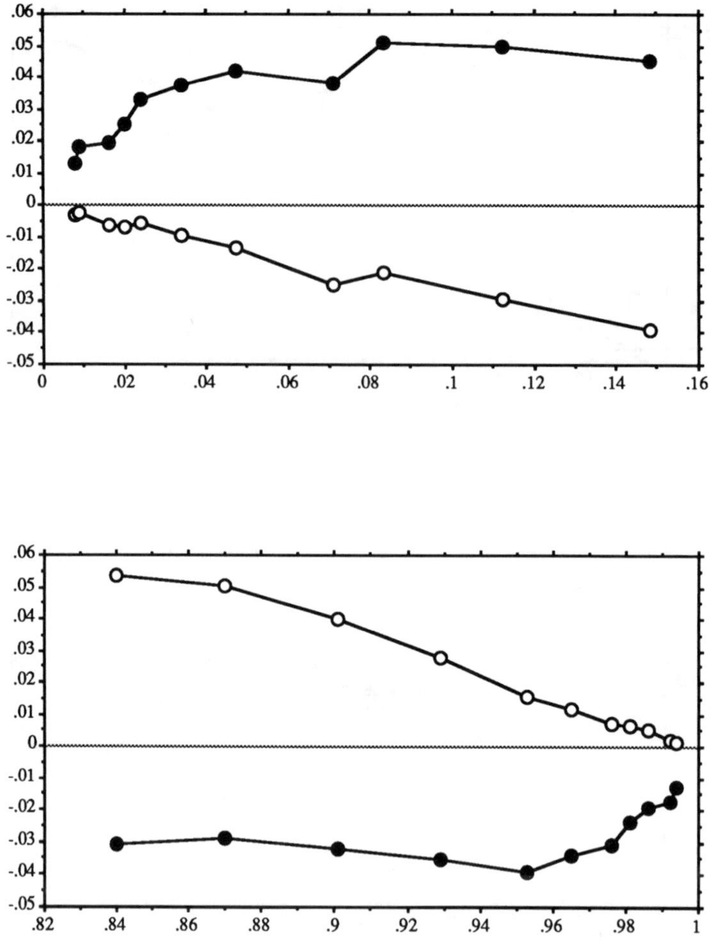

Figure 4b. *Monte Carlo estimates of the* **bias** *of estimates of the distribution function of a unstudentized linear contrast and a studentized linear contrast, plotted against the distribution function. (Model D,* $p = 5$, $c = c_2$ *).* ○ *bootstrap of the unstudentized linear contrast,* ● *bootstrap of the studentized statistic.*

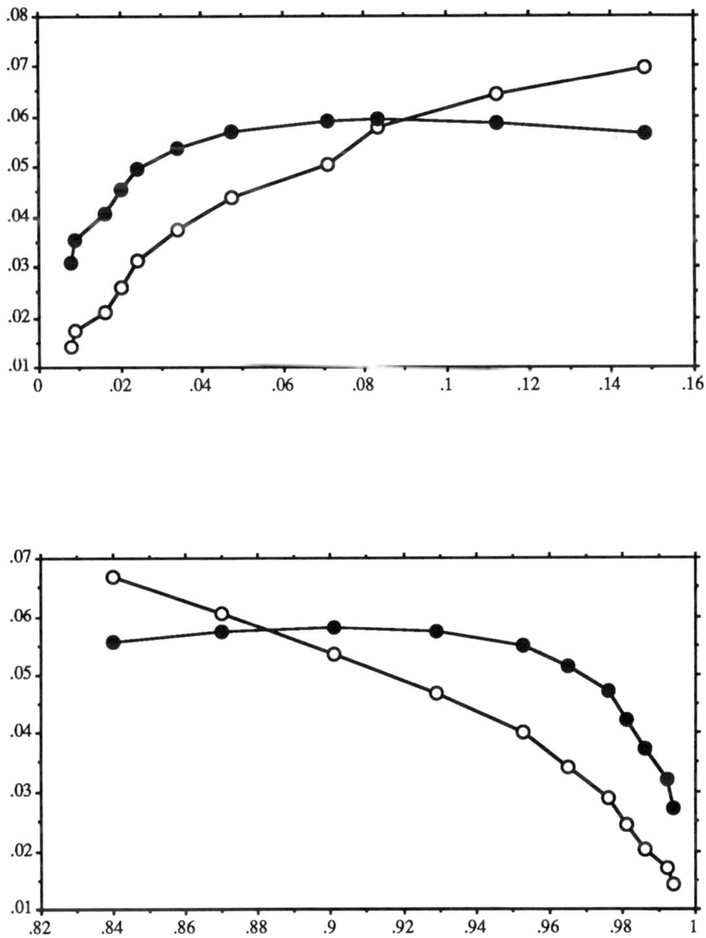

Figure 4c. *Monte Carlo estimates of the* **standard deviation** *of estimates of the distribution function of a unstudentized linear contrast and a studentized linear contrast , plotted against the distribution function. (Model D,* p = 5, c = c_2 *).* ○ *bootstrap of the unstudentized linear contrast,* ● *bootstrap of the studentized statistic.*

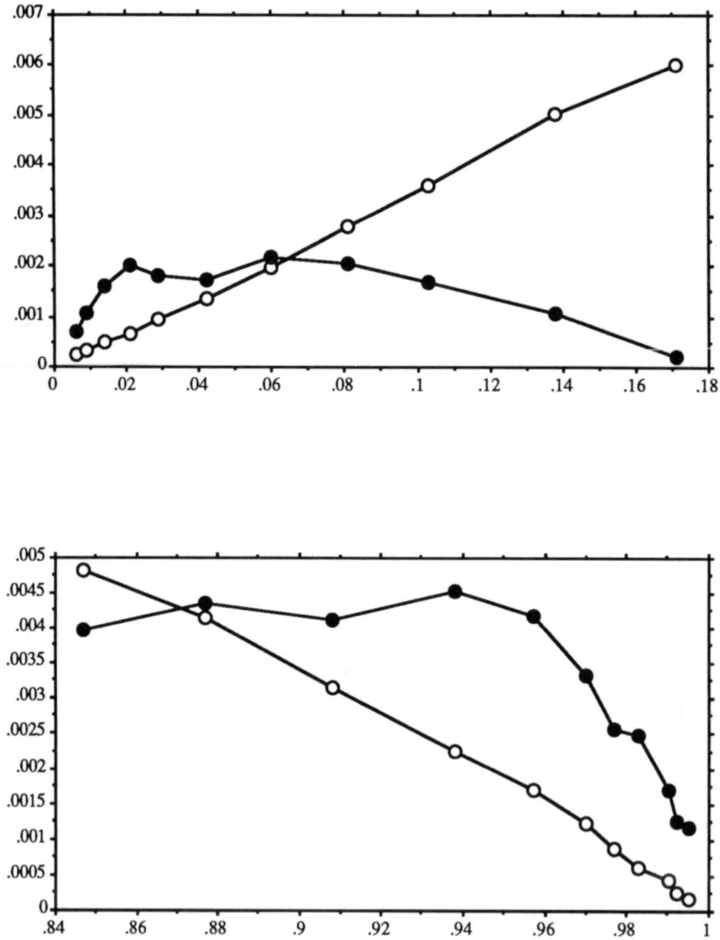

Figure 4d. *Monte Carlo estimates of the* **expected squared error** *of estimates of the distribution function of a unstudentized linear contrast and a studentized linear contrast , plotted against the distribution function. (Model D,* p = 10, c = c_2*).* ○ *bootstrap of the unstudentized linear contrast,* ● *bootstrap of the studentized statistic.*

9. References.

Abramovitch, L. and Singh, K. (1985). Edgeworth corrected pivotal statistics and the bootstrap. *Ann. Statist.* **13** 116 - 132.

Araujo, A. and Giné, E. (1980). *The central limit theorem for real and banach valued random variables*. Wiley, New York.

Arcones, M. A. and Giné, E. (1990). Some bootstrap tests of symmetry for univariate continous distributions. *Ann. Statist.* **19** 1496 - 1511.

Athreya, K.B. (1987). Bootstrap of the mean in the infinite variance case. *Ann. Statist.* **15** 724 - 731.

Azzalini, A., Bowman, A.W., and Härdle, W. (1989). On the use of nonparametric regression for model checking. *Biometrika.* **76** 1 - 11.

Babu, G. J. and Singh, K. (1984). On one - term Edgeworth correction by Efron's bootstrap. *Sankhya, Ser. A* **46** 219 - 232.

Basawa, I. V., Mallik, A. K., McCormick, W. P., Reeves, J. H., and Taylor, R. L. (1991). Bootstrapping unstable first - order autoregressive processes. *Ann. Statist.* **19** 1098 - 1101.

Basawa, I. V., Mallik, A. K., McCormick, W. P., and Taylor, R. L. (1989). Bootstrapping explosive autoregressive processes. *Ann. Statist.* **17** 1479 - 1486.

Beran, R. (1982). Estimated sampling distributions: The bootstrap and competitors. *Ann. Statist.* **10** 212 - 225 .

Beran, R. (1984a). Jackknife approximation to bootstrap estimates. *Ann. Statist.* **12** 101 - 118.

Beran, R. (1984b). Bootstrap methods in statistics. *Jahresber. Math. Ver.* **86** 14-30.

Beran, R. (1986). Discussion to Wu,C.F.J.: Jackknife, bootstrap and other resampling methods in regression analysis. *Ann. Statist.* **14** 1295-1298.

Beran, R. (1987). Prepivoting to reduce level error of confidence sets. *Biometrika* **74** 457 - 468.

Beran, R. (1988). Prepivoting test statistics : A bootstrap view of asymptotic refinements. *J. Amer. Statist. Assoc.* **83** 687 - 697.

Beran, R. and Ducharme, G. R. (1991). *Asymptotic Theory for Bootstrap Methods in Statistics*. Publications CRM, Universite de Montreal

Beran, R. and Srivastava, M. S. (1985). Bootstrap tests and confidence regions for functions of a covariance matrix. *Ann. Statist.* **13** 95 - 115. [Correction: **15** (1987) 470 - 471.]

Bhattacharya, R. (1987). Some aspects of Edgeworth expansions in statistics and probability. In *New Perspectives in Theoretical and Applied Statistics* (M. L. Puri, J. P. Vilaplana and W. Wertz, eds.) 157 - 171 Wiley, New York.

Bhattacharya, R. and Qumsiyeh, M. (1989). Second order and L^p - comparisons between the bootstrap and empirical Edgeworth expansion methologies. *Ann. Statist.* **17** 160 169.

Bickel, P. J. and Freedman, D. A. (1980). On Edgeworth expansions for the bootstrap. *Technical report.* Department of Statistics, University of California, Berkeley.

Bickel, P. J. and Freedman, D. A. (1981). Some asymptotic theory for the bootstrap. *Ann. Statist.* **9** 1196 - 1217.

Bickel, P. J. and Freedman, D. A. (1983). Bootstrapping regression models with many parameters. *A Festschrift for Erich Lehmann* (P. Bickel, K. Doksum, and J. C. Hodges, eds.) 28-48. Wadsworth, Belmont, California.

Bickel, P. J. and Rosenblatt, M. (1973). On some global measures of the deviations of density function estimates. *Ann. Statist.* **1** 1071 -1095. [Correction: **3** (1975) 1370.]

Billingsley, P. (1968). *Convergence of probability measures*. Wiley, New York.

Bose, A. (1988). Edgeworth correction by bootstrap in autoregressions. *Ann. Statist.* **16** 1709 - 1722.

Bretagnolle, J. (1983). Lois limites du bootstrap de certaines fonctionelles. *Ann. Inst. H. Poincaré, Sec. B* **19** 281 - 296.

Cao - Abad, R. (1991). Rate of convergence for the wild bootstrap in nonparametric regression. *Ann. Statist.* **19** 2226 - 2231.

Cao - Abad, R. and Gonzales - Manteiga, W. (1990). Bootstrap methods in regression smoothing: an alternative procedure to the wild resampling plan. *Unpublished manuscript.*

Carroll, R. J. (1982). Adapting for Heteroscedasticity in Linear Models. *Ann. Statist.* **10** 1224 - 1233.

Collomb, G. and Härdle, W. (1986). Strong Uniform Convergence Rates in Robust Nonparametric Time Series Analysis and Prediction; Kernel Regression Estimation from Dependent Observations. *Stochastic Processes and their Applications* **23** 77 - 89.

Cox, D., Koh, E., Wahba, G. and Yandell, B. (1988). Testing the (parametric) null model hypothesis in (semiparametric) partial and generalized spline models. *Ann. of Statist.* **16** 113 - 119.

Cox, D. and Koh, E. (1989). A smoothing spline based test of model adequacy in polynomial regression. *Ann. Inst. Statist. Math.* **41** 383 - 400.

Cramér, H. and Leadbetter, M. R. (1967). *Stationary and Related Processes.* Wiley, New York, London, Sydney.

Csörgö, S. and Mason, D. M. (1989). Bootstrapping empirical functions. *Ann. Statist.* **17** 1447 - 1471.

Cuevas, A. and Gonzalez Manteiga, W. (1990). Data - driven smoothing based on convexity properties . *Unpublished manuscript.*

Cuevas, A. and Gonzalez Manteiga, W. (1991). Data - driven smoothing based on convexity properties . In *Nonparametric Functional Estimation and Related Topics. Proceedings of the NATO Advanced Study Institute, Spetses (Greece)* (G. Roussas ed.) 225 - 240. Kluver, Dordrecht.

Davies, L. (1990). The asymptotics of S - estimators in the linear regression model. *Ann. Statist.* **18** 1651 - 1675.

Diciccio, T. J. and Efron, B. (1990). Better approximate confidence intervals in exponential families. *Technical report,* Stanford University.

Diciccio, T. J. and Romano, J. P.(1988a). A review of bootstrap confidence intervals (with discussion). *J. R. Statist. B* **50** 338 - 354.

Diciccio, T. J. and Romano, J. P.(1988b). On parametric bootstrap procedures for second - order accurate confidence limits. *Technical report,* Stanford University.

Diciccio, T. J. and Tibshirani, R. (1987). Bootstrap confidence intervals and bootstrap approximations. *J. Amer. Statist. Assoc.* **82** 163 - 170.

Dikta, G. (1988). Approximation of nearest neighbour regression function estimators. *Technical report, University of Gießen.*

Do, K. - A. and Hall, P. (1991). On importance resampling for the bootstrap. *Biometrika* **78** 161 - 167.

Dunford, N. and Schwartz, J. (1985). *Linear operators.* Interscience **1**, New York.

Dümbgen, L. (1991a). The asymptotic behaviour of some nonparametric change - point estimators. *Ann. of Statist.* **19** 1471 - 1496.

Dümbgen, L. (1991b). On nondifferentiable functions and the bootstrap. *Preprint.*

Efron, B.(1979). Bootstrap methods:Another look at the Jackknife. *Ann. of Statist.***7** 1 - 26.

Efron, B. (1982). *The jackknife, the bootstrap and other resampling plans.* CBMS-NSF Regional Conference Series in Applied Mathem. Monogr. **38**. SIAM, Philadelphia.

Efron, B. (1987). Better bootstrap confidence intervals. (with discussion) *J. Amer. Statist. Assoc.* **82** 171 - 200.

Efron, B. (1990). More efficient bootstrap computations. *J. Amer. Statist. Assoc.* **85** 79 - 89.

Efron, B. and Gong, G. (1983). A leisurely look at the bootstrap, the jackknife and croosvalidation. *Amer. Statistician* **37** 36-48.

Efron, B. and Stein, C. (1981). The jackknife estimate of variance. *Ann. Statist.* **9** 586 - 596.

Efron, B. and Tibshirani, R. (1986). Bootstrap methods for standard errors, confidence intervals, and other measures of statistical accuracy. *Statist. Science* **1** 54-77.

Ehm, W. (1986). On maximum likelihood estimation in high - dimensional log - linear type models. I. The independent case. *Preprint.* Sonderforschungsbereich 123, Universität Heidelberg.

Ehm, W. (1991). *Statistical problems with many parameters: critical quantities for approximate normality and posterior density based inference.* Habilitationsschrift, Universität Heidelberg.

Esseen, C. G. (1966). On the Kolmogorov-Rogozin inequality for the concentration function. *Z. Wahrscheinlichkeitstheorie verw. Geb.* **5** 210 - 216.

Eubank, R. and Spiegelman, C. (1989). Testing the goodness - of - fit of linear models via nonparametric regression techniques. *Unpublished manuscript.*

Falk, M. and Kaufmann, E. (1991). Coverage probabilities of bootstrap - confidence intervals for quantiles. *Ann. Statist..* **19** 485 - 495.

Falk, M. and Reiss, R. - D. (1989a). Weak convergence of smoothed and nonsmoothed bootstrap quantile estimates. *Ann. Probab.* **17** 362 - 371.

Falk, M. and Reiss, R. - D. (1989b). Bootstrapping the distance between smooth bootstrap and sample quantile distribution. *Probab. Theory Rel. Fields* **82** 177 - 186.

Feller, W. (1966). *An introduction to probability theory and its applications.* Vol. II. Wiley, New York.

Firth, D., Glosup, J., and Hinkley, D.V. (1989). Nonparametric curves for checking model fit. *Unpublished manuscript.*

Fisher, N.I., Mammen, E. and Marron, J.S. (1990). Testing for multimodality. *Preprint.* Sonderforschungsbereich 123, Universität Heidelberg.

Franke, J.(1990). Bootstrapping of nonlinear autoregressive time series. Some preliminary remarks. *Proceedings of the Bootstrap Conference in Trier, Germany.* Springer, Heidelberg. To appear.

Franke, J. and Härdle, W.(1990). On bootstrapping kernel spectral estimates. *Unpublished manuscript.*

Freedman,D.A.(1981). Bootstrapping regression models. *Ann of Statist.* **9** 1218-1228.

Ghosh, B. K. and Huang, W. - M. (1991). The power and optimal kernel of the Bickel - Rosenblatt test for goodness of fit. *Ann. Statist.* **19** 999 - 1009.

Gill, R. D. (1989). Non- and semi - parametric maximum likelihood estimators and the von Mises method (part 1). *Scand. J. Statist.* **16** 97 - 128.

Giné, E. and J. Zinn (1989). Necessary conditions for the bootstrap of the mean. *Ann. Statist.* **17** 684 - 691.

Giné, E. and J. Zinn (1990). Bootstrapping general empirical measures. *Ann. Prob.* **18**.

Gnedenko, B. V. and Kolmogorov, A. N. (1954). *Limit distributions for sums of independent random variables.* Addison - Wesley, Reading, Massachusetts.

Good, I. J. and Gaskins, R. A. (1980). Density estimation and bump - hunting by the penalized likelihood method exemplified by scattering and meteorite data. *J. Amer. Statist. Assoc.* **75** 42 - 73.

Götze, F. (1984). Expansions for von Mises functionals. *Z. Wahrsch. verw. Gebiete* **65** 599 - 625.

Götze, F. (1985). Asymptotic expansions in functional limit theorems. *J. Multivariate Anal.* **16** 1 - 20.

Götze, F. (1989). Edgeworth expansions in functional limit theorems. *Ann. Probab.* **17** 1602 - 1634 .

Haberman, S. J. (1977a). Log - linear and frequency tables with small expected cell counts. *Ann. Statist.* **5** 1148 - 1169 .

Haberman, S. J. (1977b), Maximum likelihood estimates in exponential response models. *Ann. Statist.* **5** 815 - 841 .

Hall, P. (1984). Central limit theorems for integrated square error of multivariate nonparametric density estimators. *J. Multivar. Anal.* **14** 1 - 16.

Hall, P. (1986a). On the bootstrap and confidence intervals. *Ann. Statist.* **14** 1431 - 1452.

Hall, P. (1986b). On the number of bootstrap simulations required to construct a confidence interval. *Ann. Statist.* **14** 1453 - 1462 .

Hall, P. (1988). Theoretical comparison of bootstrap confidence intervals. *Ann. Statist.* **16** 927 - 953.

Hall, P. (1990a). Edgeworth expansions for nonparametric density estimators, with applications. *Unpublished manuscript.*

Hall, P. (1990b). Using the bootstrap to estimate mean squared error and select smoothing parameter in nonparametric problems. *J. Multivar. Anal.* **32** 177 - 203.

Hall, P. (1990c). On bootstrap confidence intervals in nonparametric regression. *Unpublished manuscript.*

Hall, P. (1990d). Asymptotic properties of the bootstrap for heavy - tailed distributions. *Ann. Probab.* **18** 1342 - 1360.

Hall, P. (1991). Bahadur representations for uniform resampling and importance resampling with applications toasymptotic relative efficiency. *Ann. Statist.* **19** 1062 - 1072.

Hall, P. (1992). *The bootstrap and Edgeworth expansions.* Springer, New York.

Hall, P. and Marron, J. S. (1987). Extent to which least - squares cross - validation minimises integrated squared error in nonparametric density estimation. *Probab. Theory Rel. Fields* **74** 567 - 581.

Hall, P. and Marron, J. S. (1991). Lower bounds for bandwidth selection in density estimation. *Probab. Theory Rel. Fields.* To appear.

Hall, P., Marron, J. S., Park, B. U. (1989). Smoothed cross - validation. *Unpublished manuscript.*

Hall, P. and Martin, M. A. (1988a). Exact convergence rate of bootstrap quantile variance estimator. *Probab. Theory Rel. Fields* **80** 261 - 269.

Hall, P. and Martin, M. A. (1988b). On bootstrap resampling and iteration. *Biometrika* **75** 661 - 671.

Hall, P., Sheather, S. J., Jones, M. C. and Marron, J. S. (1991). On optimal data-based bandwidth selection in kernel density estimation. *Biometrika* **78** 263 - 269.

Hall, P. and Titterington, D. M. (1989). The effect of simulation order on level accuracy and power of Monte Carlo tests. *J. R. Statist. Soc.* B **51** 459 - 467.

Hampel, F. R., Ronchetti, E. M. , Rousseeuw, P. J. and Stahel, W. A. (1986). *Robust statistics. The approach based on influence functions.* Wiley, New York.

Härdle, W. (1990). *Applied nonparametric regression.* Econometric Society Monograph Series, Cambridge University Press.

Härdle, W. and Bowman, A. (1988). Bootstrapping in nonparametric regression: local adaptive smoothing and confidence bands. *J. Amer. Statist. Assoc.* **83** 102 - 110.

Härdle, W. and Mammen, E. (1990). Comparing non parametric versus parametric regression fits. *Preprint.*Sonderforschungsbereich 123, Universität Heidelberg.

Härdle, W. and Mammen, E. (1991). Bootstrap methods in nonparametric regression. In *Nonparametric Functional Estimation and Related Topics. Proceedings of the NATO Advanced Study Institute, Spetses (Greece)* (G. Roussas ed.) 111 - 124. Kluver, Dordrecht.

Härdle, W. and Marron, J. S. (1990). Semiparametric comparison of regression curves. *Ann. Statist.* **18** 63 - 89.

Härdle, W. and Marron, J. S. (1991). Bootstrap simultaneous error bars for nonparametric regression. *Ann. Statist.* **19** 778 - 796.

Hartigan, J. A. and Hartigan, P. M. (1985). The DIP test of unimodality. *Ann. Statist.* **13** 70-84.

Hartigan, J. A. (1986). Comment to Efron, B. and Tibshirani, R.: Bootstrap methods for standard errors, confidence intervals, and other measures of statistical accuracy. *Statist. Science* **1** 75 -76.

Helmers, R. (1991a). On the Edgeworth expansion and the bootstrap approximation for a studentised U - statistic. *Ann. Statist.* **19** 470 - 484.

Helmers, R. (1991b). *Bootstrap methods*. Unpublished book manuscript.

Hinkley, D. V. (1988). Bootstrap methods (with discussion). *J. R. Statist. B* **50** 321 - 337 .

Huber, P. J. (1981). *Robust statistics*. Wiley, New York.

Janas, D. (1991). A smoothed bootstrap estimator for a studentized sample quantile. *Preprint*. Sonderforschungsbereich 123, Universität Heidelberg.

Jeganathan, P. (1990). On Edgeworth expansions for estimators in time series models. I. Nonlinear regression models. *Unpublished manuscript.*

Joeckel, K. H., Rothe, G., and Sendler, W. (eds.) (1992). *Bootstrapping and related techniques*. Lecture Notes in Econometrics and Math. Systems Vol. 376. Springer, New York.

Johns, M. V. (1988). Importance sampling for bootstrap confidence intervals. *J. Amer. Statist. Assoc.* **83** 709 - 714.

Jong, P. de (1987). A central limit theorem for generalized quadratic forms. *Probab. Th. Rel. Fields* **75** 261 - 277.

Karlin, S. and Rinott, Y. (1982). Applications of ANOVA type decompositions of conditional variance statistics including Jackknife estimates. *Ann. Statist.* **10** 485 - 501.

Kantorowitsch, L. W. and Akilow, G. P. (1964). *Funktionalanalysis in normierten Räumen.*. Akademie Verlag, Berlin.

Knight, K. (1989). On the bootstrap of the sample mean in the infinite variance case. *Ann. Statist.* **17** 1168 - 1175 .

Koehler, K. J. (1986). Goodness - of - fit tests for log-linear models in sparse contingency tables. *J. Amer. Statist. Assoc.* **81** 483 - 493.

Kreiss, J. - P. (1988). Bootstrap procedures for AR(∞) - processes. *Unpublished manuscript.*

Künsch, H. R. (1989). The jackknife and the bootstrap for general stationary observations. *Ann. Statist.* **17** 1217 - 1241.

Le Cam, L. (1986). *Asymptotic methods in statistical decision theory.* Springer, New York.

LePage, R. and Billard, L. (eds.) (1992). *Exploring the limits of bootstrap.* Wiley, New York.

Liu, R. (1988). Bootstrap procedures under some non i.i.d. models. *Ann. Stat.* **16** 1696 - 1708.

Liu, R. and Singh K. (1987). On a partial correction by the bootstrap . *Ann. Stat.* **15** 1713 - 1718.

Liu, R. and Singh, K. (1989). Efficiency and robustness in resampling. *Unpublished manuscript.*

Liu, R. and Singh, K. (1991). Using i.i.d. bootstrap for general non - i.i.d. models. *Unpublished manuscript.*

Mack, Y.P. and Silverman, B.W. (1982). Weak and strong uniform consistency of kernel regression estimates. *Zeitschrift für Wahrscheinlichkeitstheorie und verwandte Gebiete* **61** 405 - 415.

Mammen, E. (1989a). Asymptotics with increasing dimension for robust regression with applications to the bootstrap. *Ann. Statist.* **17** 382 - 400.

Mammen, E. (1989b). Bootstrap and wild bootstrap for high - dimensional linear models. *Preprint.* Sonderforschungsbereich 123, Universität Heidelberg.

Mammen, E. (1990). Higher - order accuracy of bootstrap for smooth functionals. *Preprint.* Sonderforschungsbereich 123, Universität Heidelberg.

Mammen, E. (1991a). Estimating a smooth monotone regression function. *Ann. Statist.* **19** 724 - 740.

Mammen, E. (1991b). Nonparametric regression under qualitative smoothness assumptions. *Ann. Statist.* **19** 741 - 759.

Mammen, E. (1991c). Nonparametric curve estimation and simple curve characteristics. In *Nonparametric Functional Estimation and Related Topics. Proceedings of the NATO Advanced Study Institute, Spetses (Greece)* (G. Roussas ed.) 133 - 140. Kluver, Dordrecht.

Mammen, E. (1991d). On qualitative smoothness of kernel density estimates. *Preprint.* Sonderforschungsbereich 123, Universität Heidelberg.

Mammen, E. (1992). Bootstrap, wild bootstrap, and asymptotic normality. *Preprint.* Sonderforschungsbereich 123, Universität Heidelberg.

Mammen, E. , Marron, J. S. and Fisher,N. I. (1992). Some asymptotics for multimodality tests based on kernel density estimates. *Probab. Theory Rel. Fields.* In press.

Martin, M. A. (1990). On bootstrap iteration for coverage correction in confidence intervals. *J. Statist. Amer. Assoc.* **85** 1105 - 1118.

McDonald, J. A. (1982). Projection pursuit regression with the ORION I workstation. *20 minute, 16 mm color sound film.* Stanford University.

Müller, D.W. and Sawitzki, G. (1990). Using excess mass estimates to investigate the modality of a distribution. *Proceedings of the ICOSCO - I Conference (First International Conference on Statistical Computing, Çesme, Izmir, Turkey, 1987). Ed. by E. C. van der Meulen.*

Müller, D. W. and Sawitzki, G. (1991). Excess mass estimates and tests for multimodality. *J. Amer. Statist. Assoc.* **86** 738 - 746.

Nadaraya, E.A. (1964). On estimating regression. *Theory Prob. Appl.* **10** 186 - 190.

Navidi, W. (1989). Edgeworth expansions for bootstrapping regression models. *Ann. Statist.* **17** 1472 - 1478.

Neuhaus, G. (1986). A class of quadratic goodness - of - fit tests. *Unpublished manuscript.*

Neuhaus, G. (1988). Addendum to "Local Asymptotics for Linear Rank Statistics with estimated Score Functions" . *Ann. Statist.* **16** 1342 - 1343.

Park, B.U. and Marron, J.S. (1990). Comparison of data - driven bandwidth selectors. *J. Amer. Statist. Assoc.* **85** 66 .

Parr, W. C. (1985). Jackknifing differentiable statistical functionals. *J. Roy. Statist. Soc. Ser. B* **47** 56 - 66.

Pflanzagl, J. and Wefelmeyer W. (1985). Asymptotic expansions for general statistical models. *Lecture Notes in Statistics* **31**, Springer, Berlin, Heidelberg.

Portnoy, S. (1984). Asymptotic behaviour of M-estimators of p regression parameters when $\frac{p^2}{n}$ is large. I. Consistency. *Ann. Statist.* **12** 1298 - 1309.

Portnoy, S. (1985). Asymptotic behaviour of M-estimator of p regression parameters when $\frac{p^2}{n}$ is large. II. Normal approximation. *Ann. Statist.* **13** 1403 - 1417. [Correction: **19** (1991) 2282.]

Portnoy, S. (1988). Asymptotic behaviour of likelihood methods for exponential families when the number of parameters tends to infinity. *Ann. Statist.* **16** 356 - 366.

Raikov, D. A. (1938). On a connection between the central limit theoremin the theory of probability and the law of large numbers. *Izvestiya Akad. Nauk SSR, Ser. Mat.* 323 - 338.

Raz, J. (1990). Testing for no effect when estimating a smooth function by nonparametric regression: a randomization approach. *J. Amer. Statist. Assoc.* **85** 132 - 138.

Reeds, J. A. (1976). *On the definition of von Mises functionals.* Thesis. Harvard University. Cambridge, Massachusetts.

Romano, J. H. (1988). A bootstrap revival of some non - parametric distance tests. *J. Amer. Statist. Assoc.* **83** 698 - 708.

Romano, J. H. (1989). Bootstrap and randomization tests of some nonparametric hypotheses. *Ann. Statist.* **14** 141 - 159.

Rousseeuw, P. J. and Yohai, V. J. (1984). Robust regression by means of S-estimators. *Robust and Nonlinear Time Series Analysis* (J. Franke, Härdle, W. and Martin, D., eds.). *Lecture Notes in Statistics* **26** 256 - 272. Springer, New York .

Sauermann, W. (1986). Bootstrap - Verfahren in log - linearen Modellen. Dissertation. Universität Heidelberg.

Sauermann, W. (1989). Bootstrapping the maximum likelihood estimator in high-dimensional log-linear models. *Ann. Statist.* **17** 1198 - 1216.

Sen, P. K. (1988a). Functional jackknifing: Rationality and general asymptotics. *Ann. Statist.* **16** 450 - 469.

Sen, P. K. (1988b). Functional approaches in resampling plans: a review of some recent developments. *Sankhya* **50** 394 - 435.

Shao, J. (1990). Bootstrap estimation of the asymptotic variances of statistical functionals. *Ann. Inst. Statist. Math.* **42** 737 - 752.

Shao, J. and Wu, C. F. J. (1989). A general theory for jackknife variance estimation. *Ann. Statist.* **17** 1176 - 1197.

Shorack, G. (1982). Bootstrapping robust regression. *Comm. Statist.* **A 11** 961 - 972.

Silverman, B. W. (1978). Weak and strong uniform consistency of the kernel estimate of a density and its derivatives. *Ann. Statist.* **6** 177 - 184.

Silverman, B. W. (1981). Using kernel estimates to investigate multimodality. *Journal of the Royal Statistical Society, Series B* **43** 97 - 99.

Silverman, B. W. (1983). Some properties of a test for multimodality based on kernel density estimates. *Probability, Statistics and Analysis* (Kingman, J.F.C. and Reuter, G.E.H. ,eds.) 248 - 259. Cambridge University Press, Cambridge.

Silverman, B. W. (1986). *Density estimation for statistics and data analysis.* Chapman and Hall, London.

Singh, K. (1981). On the asymptotic accuracy of Efron's bootstrap. *Ann. Statist.* **9** 1187-1195.

van Zwet, W. R. (1984). A Berry - Esséen bound for symmetric statistics. *Z. Wahrsch. Verw. Gebiete* **66** 425 - 440.

van Zwet, W. (1989). Hoeffding's decomposition and the bootstrap. Talk given at the conference on "Asymptotic methods for computer-intensive procedures in statistics" in Oberwolfach, West-Germany.

Watson, G. S. (1964). Smooth regression analysis. *Sankhya, Series A* 359 - 372

Wu, C. F. J. (1986). Jackknife, bootstrap, and other resampling methods in regression analysis (with discussion). *Ann. Statist.* **14** 1261-1295.

Yin,Y. Q., Bai, Z. D., and Krishnaiah, P. R. (1988). On the limit of the largest eigenvalue of the large dimensional sample covariance matrix. *Probab. Th. Rel. Fields* **78** 509 - 521.

Yohai, V. J. (1987). High breakdown point and high efficiency robust estimates for regression. *Ann. Statist.* **15** 642 - 656.

Vol. 1: R.A. Fisher: An Appreciation. Edited by S.E. Fienberg and D.V. Hinkley. XI, 208 pages, 1980.

Vol. 2: Mathematical Statistics and Probability Theory. Proceedings 1978. Edited by W. Klonecki, A. Kozek, and J. Rosinski. XXIV, 373 pages, 1980.

Vol. 3: B.D. Spencer, Benefit-Cost Analysis of Data Used to Allocate Funds. VIII, 296 pages, 1980.

Vol. 4: E.A. van Doorn, Stochastic Monotonicity and Queueing Applications of Birth-Death Processes. VI, 118 pages, 1981.

Vol. 5: T. Rolski, Stationary Random Processes Associated with Point Processes. VI, 139 pages, 1981.

Vol. 6: S.S. Gupta and D.-Y. Huang, Multiple Statistical Decision Theory: Recent Developments. VIII, 104 pages, 1981.

Vol. 7: M. Akahira and K. Takeuchi, Asymptotic Efficiency of Statistical Estimators. VIII, 242 pages, 1981.

Vol. 8: The First Pannonian Symposium on Mathematical Statistics. Edited by P. Révész, L. Schmetterer, and V.M. Zolotarev. VI, 308 pages, 1981.

Vol. 9: B. Jørgensen, Statistical Properties of the Generalized Inverse Gaussian Distribution. VI, 188 pages, 1981.

Vol. 10: A.A. McIntosh, Fitting Linear Models: An Application of Conjugate Gradient Algorithms. VI, 200 pages, 1982.

Vol. 11: D.F. Nicholls and B.G. Quinn, Random Coefficient Autoregressive Models: An Introduction. V, 154 pages, 1982.

Vol. 12: M. Jacobsen, Statistical Analysis of Counting Processes. VII, 226 pages, 1982.

Vol. 13: J. Pfanzagl (with the assistance of W. Wefelmeyer), Contributions to a General Asymptotic Statistical Theory. VII, 315 pages, 1982.

Vol. 14: GLIM 82: Proceedings of the International Conference on Generalised Linear Models. Edited by R. Gilchrist. V, 188 pages, 1982.

Vol. 15: K.R.W. Brewer and M. Hanif, Sampling with Unequal Probabilities. IX, 164 pages, 1983.

Vol. 16: Specifying Statistical Models: From Parametric to Non-Parametric, Using Bayesian or Non-Bayesian Approaches. Edited by J.P. Florens, M. Mouchart, J.P. Raoult, L. Simar, and A.F.M. Smith, XI, 204 pages, 1983.

Vol. 17: I.V. Basawa and D.J. Scott, Asymptotic Optimal Inference for Non-Ergodic Models. IX, 170 pages, 1983.

Vol. 18: W. Britton, Conjugate Duality and the Exponential Fourier Spectrum. V, 226 pages, 1983.

Vol. 19: L. Fernholz, von Mises Calculus For Statistical Functionals. VIII, 124 pages, 1983.

Vol. 20: Mathematical Learning Models — Theory and Algorithms: Proceedings of a Conference. Edited by U. Herkenrath, D. Kalin, W. Vogel. XIV, 226 pages, 1983.

Vol. 21: H. Tong, Threshold Models in Non-linear Time Series Analysis. X, 323 pages, 1983.

Vol. 22: S. Johansen, Functional Relations, Random Coefficients and Nonlinear Regression with Application to Kinetic Data, VIII, 126 pages, 1984.

Vol. 23: D.G. Saphire, Estimation of Victimization Prevalence Using Data from the National Crime Survey. V, 165 pages, 1984.

Vol. 24: T.S. Rao, M.M. Gabr, An Introduction to Bispectral Analysis and Bilinear Time Series Models. VIII, 280 pages, 1984.

Vol. 25: Time Series Analysis of Irregularly Observed Data. Proceedings, 1983. Edited by E. Parzen. VII, 363 pages, 1984.

Vol. 26: Robust and Nonlinear Time Series Analysis. Proceedings, 1983. Edited by J. Franke, W. Härdle and D. Martin. IX, 286 pages, 1984.

Vol. 27: A. Janssen, H. Milbrodt, H. Strasser, Infinitely Divisible Statistical Experiments. VI, 163 pages, 1985.

Vol. 28: S. Amari, Differential-Geometrical Methods in Statistics. V, 290 pages, 1985.

Vol. 29: Statistics in Ornithology. Edited by B.J.T. Morgan and P.M. North. XXV, 418 pages, 1985.

Vol 30: J. Grandell, Stochastic Models of Air Pollutant Concentration. V, 110 pages, 1985.

Vol. 31: J. Pfanzagl, Asymptotic Expansions for General Statistical Models. VII, 505 pages, 1985.

Vol. 32: Generalized Linear Models. Proceedings, 1985. Edited by R. Gilchrist, B. Francis and J. Whittaker. VI, 178 pages, 1985.

Vol. 33: M. Csörgo, S. Csörgo, L. Horváth, An Asymptotic Theory for Empirical Reliability and Concentration Processes. V, 171 pages, 1986.

Vol. 34: D.E. Critchlow, Metric Methods for Analyzing Partially Ranked Data. X, 216 pages, 1985.

Vol. 35: Linear Statistical Inference. Proceedings, 1984. Edited by T. Calinski and W. Klonecki. VI, 318 pages, 1985.

Vol. 36: B. Matérn, Spatial Variation. Second Edition. 151 pages, 1986.

Vol. 37: Advances in Order Restricted Statistical Inference. Proceedings, 1985. Edited by R. Dykstra, T. Robertson and F.T. Wright. VIII, 295 pages, 1986.

Vol. 38: Survey Research Designs: Towards a Better Understanding of Their Costs and Benefits. Edited by R.W. Pearson and R.F. Boruch. V, 129 pages, 1986.

Vol. 39: J.D. Malley, Optimal Unbiased Estimation of Variance Components. IX, 146 pages, 1986.

Vol. 40: H.R. Lerche, Boundary Crossing of Brownian Motion. V, 142 pages, 1986.

Vol. 41: F. Baccelli, P. Brémaud, Palm Probabilities and Stationary Queues. VII, 106 pages, 1987.

Vol. 42: S. Kullback, J.C. Keegel, J.H. Kullback, Topics in Statistical Information Theory. IX, 158 pages, 1987.

Vol. 43: B.C. Arnold, Majorization and the Lorenz Order: A Brief Introduction. VI, 122 pages, 1987.

Vol. 44: D.L. McLeish, Christopher G. Small, The Theory and Applications of Statistical Inference Functions. VI, 124 pages, 1987.

Vol. 45: J.K. Ghosh (Ed.), Statistical Information and Likelihood. 384 pages, 1988.

Vol. 46: H.-G. Müller, Nonparametric Regression Analysis of Longitudinal Data. VI, 199 pages, 1988.

Vol. 47: A.J. Getson, F.C. Hsuan, {2}-Inverses and Their Statistical Application. VIII, 110 pages, 1988.

Vol. 48: G.L. Bretthorst, Bayesian Spectrum Analysis and Parameter Estimation. XII, 209 pages, 1988.

Vol. 49: S.L. Lauritzen, Extremal Families and Systems of Sufficient Statistics. XV, 268 pages, 1988.

Vol. 50: O.E. Barndorff-Nielsen, Parametric Statistical Models and Likelihood. VII, 276 pages, 1988.

Vol. 51: J. Hüsler, R.-D. Reiss (Eds.), Extreme Value Theory. Proceedings, 1987. X, 279 pages, 1989.

Vol. 52: P.K. Goel, T. Ramalingam, The Matching Methodology: Some Statistical Properties. VIII, 152 pages, 1989.

Vol. 53: B.C. Arnold, N. Balakrishnan, Relations, Bounds and Approximations for Order Statistics. IX, 173 pages, 1989.

Vol. 54: K.R. Shah, B.K. Sinha, Theory of Optimal Designs. VIII, 171 pages, 1989.

Vol. 55: L. McDonald, B. Manly, J. Lockwood, J. Logan (Eds.), Estimation and Analysis of Insect Populations. Proceedings, 1988. XIV, 492 pages, 1989.

Vol. 56: J.K. Lindsey, The Analysis of Categorical Data Using GLIM. V, 168 pages, 1989.

Vol. 57: A. Decarli, B.J. Francis, R. Gilchrist, G.U.H. Seeber (Eds.), Statistical Modelling. Proceedings, 1989. IX, 343 pages, 1989.

Vol. 58: O.E. Barndorff-Nielsen, P. Blæsild, P.S. Eriksen, Decomposition and Invariance of Measures, and Statistical Transformation Models. V, 147 pages, 1989.

Vol. 59: S. Gupta, R. Mukerjee, A Calculus for Factorial Arrangements. VI, 126 pages, 1989.

Vol. 60: L. Györfi, W. Härdle, P. Sarda, Ph. Vieu, Nonparametric Curve Estimation from Time Series. VIII, 153 pages, 1989.

Vol. 61: J. Breckling, The Analysis of Directional Time Series: Applications to Wind Speed and Direction. VIII, 238 pages, 1989.

Vol. 62: J.C. Akkerboom, Testing Problems with Linear or Angu-lar Inequality Constraints. XII, 291 pages, 1990.

Vol. 63: J. Pfanzagl, Estimation in Semiparametric Models: Som Recent Developments. III, 112 pages, 1990.

Vol. 64: S. Gabler, Minimax Solutions in Sampling from Fini Populations. V, 132 pages, 1990.

Vol. 65: A. Janssen, D.M. Mason, Non-Standard Rank Tests. V 252 pages, 1990.

Vol. 66: T. Wright, Exact Confidence Bounds when Samplir from Small Finite Universes. XVI, 431 pages, 1991.

Vol. 67: M.A. Tanner, Tools for Statistical Inference: Observe Data and Data Augmentation Methods. VI, 110 pages, 1991.

Vol. 68: M. Taniguchi, Higher Order Asymptotic Theory for Tim Series Analysis. VIII, 160 pages, 1991.

Vol. 69: N.J.D. Nagelkerke, Maximum Likelihood Estimation Functional Relationships. V, 110 pages, 1992.

Vol. 70: K. Iida, Studies on the Optimal Search Plan. VIII, 13 pages, 1992.

Vol. 71: E.M.R.A. Engel, A Road to Randomness in Physic Systems. IX, 155 pages, 1992.

Vol. 72: J.K. Lindsey, The Analysis of Stochastic Processes usir GLIM. VI, 294 pages, 1992.

Vol. 73: B.C. Arnold, E. Castillo, J.-M. Sarabia, Conditional Specified Distributions. XIII, 151 pages, 1992.

Vol. 74: P. Barone, A. Frigessi, M. Piccioni, Stochastic Mode Statistical Methods, and Algorithms in Image Analysis. VI, 25 pages, 1992.

Vol. 75: P.K. Goel, N.S. Iyengar (Eds.), Bayesian Analysis Statistics and Econometrics. XI, 410 pages, 1992.

Vol. 76: L. Bondesson, Generalized Gamma Convolutions ar Related Classes of Distributions and Densities. VIII, 173 page 1992.

Vol. 77: E. Mammen, When Does Bootstrap Work? Asymptot Results and Simulations. VI, 196 pages, 1992.